T0276824

Microwave Heating Handbook

Microwave Heating Handbook

Edited by **Doreen Rowe**

New York

Published by NY Research Press,
23 West, 55th Street, Suite 816,
New York, NY 10019, USA
www.nyresearchpress.com

Microwave Heating Handbook
Edited by Doreen Rowe

International Standard Book Number: 978-1-63238-328-0 (Hardback)

Printed in the United States of America.

Contents

Preface

This book presents the readers with all the essential aspects associated with the phenomenon of microwave heating. Microwave heating has numerous applications ranging from the microwave ovens in kitchen for heating food, to a sterilization setup for medical treatment, to processing of materials in the distinct fields. This phenomenon holds considerable benefits over the conventional techniques in the aspects of less processing time and reduced environmental impacts as reflected through those applications. This book covers both the general understandings as well as advance applications of microwave heating. The book contains discussions and information on the application of this technique in the areas of chemical engineering, forestry, food, mineral processing industry and agriculture in order to provide a base for future research. The aim of this book is to serve as an elementary reference source to help the readers focus on important aspects behind the success of microwave heating.

Significant researches are present in this book. Intensive efforts have been employed by authors to make this book an outstanding discourse. This book contains the enlightening chapters which have been written on the basis of significant researches done by the experts.

Finally, I would also like to thank all the members involved in this book for being a team and meeting all the deadlines for the submission of their respective works. I would also like to thank my friends and family for being supportive in my efforts.

Editor

Section 1

General

Chapter 1

Microwave Heating Applications in Mineral Processing

S.M. Javad Koleini and Kianoush Barani

Additional information is available at the end of the chapter

1. Introduction

1.1. History

The thermal treatment of ore to bring about thermal fracturing, and thereby a reduction in Ore strength is by no means a novel idea. The first century BC Greek historian, Diodorus Siculus, recorded in his Bibliotheca Historica the ancient practice of fire setting, verifying his work with that of another Greek historian, Agatharcides, who had visited the gold mines in Egypt around the second century BC (Meyer, 1997[1]).

Oldfather, 1967[2], provides a translation of Diodorus's account of the practice: "The gold bearing earth which is hardest they first burn with a hot fire, and when they have crumbled it...they continue the working of it by hand; and the soft rock which can yield to moderate effort is crushed with a sledge".

The practice of fire setting basically consisted of constructing a large fire against the rock face to be mined. As the rock heated unevenly, it would fracture internally, severely weakening the rock. After the fires died down the rock face would be doused with water, though whether this rapid quenching was employed to further weaken the rock or to allow the miners to immediately continue working the rock face is not known (The Tech, 1886)[3].

Using this process, it was possible to weaken the rock face to the depth of approximately a foot at a time, after which the soft ore was mined and when the harder rock face was again encountered, fire setting was again employed (Cowen, 1999)[4].

Archeological evidence supports the notion that the practice of fire-setting was a worldwide phenomenon and may indeed be much older than those activities reported in the records of Diodorus Siculus, with ancient mining sites discovered at Rudna Glava in the Balkans suggesting the use of fire setting around 4500 to 4000 BC, at Ai Bunar in southern Bulgaria also dated at several thousand years BC and from which it is estimated that between 20 000 and 30 000 tonnes of ore were mined while employing the method when required (Cowen, 1999), at the ancient mining sites around Isle Royale in the Lake Superior region in North America to mine copper and up until just a few centuries ago in Japan for creating long tunnels (The Tech, 1886)[3].

In fact, it remained a vital part of the mining industry until the first use of gunpowder for blasting in 1613 (The Tech, 1886)3, after which the use of thermal treatment declined in favor of the quicker processes of drilling and blasting.

1.2. Initial studies on minerals breakage

It is reported in a review paper by Fitzgibbon and Veasey, 1990[5], that work on the use of thermal treatment to aid in rock breakage during comminution processes began again early in the 20th century, with practical studies on Cornish tin ores (Yates, 1919) [6]and quartzites (Holman, 1927)[7]. Fitzgibbon and Veasey, 1990, report that this early work showed that the thermal pretreatment of ores before comminution resulted not only in a reduction in the strength of the ores studied, but also in fewer fines being produced. Work by Myer, 1925[8], and Holman, 1927, also studied the dependence of the susceptibility of ores to heat treatment on particle size and concluded that the effectiveness of the treatment decreased with particle size (Fitzgibbon and Veasey, 1990).

1.3. Economical evaluation

In the second half of 20 century, many researchers studied on the economical aspect of conventional heat treatment. As early as 1962, it was known that the effect of thermal treatment on ore strength varies with ore mineralogy, and that fluorites and barites, in particular, are susceptible to this effect, but studies showed that the process of thermal treatment was uneconomical when compared to the use of conventional grinding alone (Prasher, 1987)[9], due to the enormous energy requirements associated with heating the bulk ore to the required temperatures, where Wills et al., 1987[10], report that other workers have calculated that the cost of heat treatment and subsequent grinding could be as high as 6 times that of conventional grinding alone (Scheding et al., 1981)[11].

1.4. Water quenching after heat treatment

Some researchers studied on the effect of water quenching after heat treatment to reduce the economic costs of heat treatment process.

Kanellopoulos and Ball, 1975[12], studied the effect of heat treatment on crushing and grinding of quartzite samples. Their investigations showed that heat treatment above 400°C improves the comminution of the ore, but that the best results are obtained after heating the quartzite to temperatures above the α-β phase transition temperature of quartz (i.e. 573°C), at which a sudden volumetric expansion (i.e. a volume increase of 0.86%) of quartz crystals occurs. Comparative testing of material which was slow cooled from 680°C to ambient, and material which was shock cooled through water quenching, showed no difference in the product size distribution of the material after milling. Comparisons of results obtained from the same heat treatments after comminution by slow crushing, however, indicate that quenching the ore results in a change in the product particle size distribution, with significantly less material passing at larger sizes with the difference in passing size decreasing with particle size, thus resulting in a finer product without a significant increase in the production of very fine material. This was the first indication that the manner of the post-processing of the material may be as important as the thermal treatment itself.

Pocock et al., 1998[13], investigated the use of various quenching solutions to ascertain whether any improvement could be seen from using acid, alkali or salt solutions instead of water. It was found that all of these showed improvements in grinding energy reduction over the use of water, and of these, it was found that the use of acid or alkali solutions provided the best results. At the same time, it was seen from UFLC tests that as comminution of the treated particles continued (i.e. as the particles become smaller), the observed effects of the thermal pretreatment are reduced. What this indicates is that as the easily exploited newly formed fractures are used up, the strength of the ore begins to once again approach that of the untreated ore.

1.5. Minerals liberation and heat treatment

Wills et al., 1987[10], investigated the thermally assisted liberation of cassiterite in an ore mined at South Crofty. Previous work on this ore (Sherring, 1981)[14] had shown a 55% reduction in grinding resistance when the ore was heated to 650°C and then rapidly cooled, however, this was greatly offset by the energy required to heat the material. It was suggested by Manser, 1983[15], that an increase in tin recovery of 1% would offset this cost in the case of the South Crofty ore, due to the value of the recovered minerals. Employing similar conditions in their work, and heat treated polished sections of the ore which could be photographed before and after the treatment to look for any induced fractures which might indicate that this increase in liberation may be possible. Their results showed that while some intergranular fracturing was observed as a result of their heat treatment, in most of the cases extensive transgranular fracturing occurred, and later separation tests showed no enhanced liberation or recovery of this material with heat treatment.

2. Microwave treatment

2.1. Minerals in microwave field

Conventional heat treatment of minerals is a process with high-energy consumption and it is not economical. Hence, researchers, searched for processes that are more effective.

It is reported in a review paper by Xia and Pickles, 1997[16], that the earliest work on the microwaving of minerals began with a study of the high temperature processing of certain oxides and sulfides using a resonant cavity operating at 2.45 GHz and variable power up to 1.6 kW (Ford and Pei, 1967)[17]. Table 1 shows the results. The results of this early work were qualitative in nature, concluding that, in general, dark coloured compounds heated rapidly (reaching temperatures of up to 1000°C), while lighter coloured compounds heated slower but were capable of being heated to higher temperatures.

Compound	Heating time (min)	Max. Temp. (°C)
Al_2O_3	24	1900
C	0.2	1000
CaO	40	200
Co_2O_3	3	900
CuO	4	800
CuS	5	600
Fe_2O_3	6	1000
Fe_3O_4	0.5	500
FeS	6	800
MgO	40	1300
MoO_3	0.46	750
MoS_2	0.1	900
Ni_2O_3	3	1300
PbO	13	900
UO_2	0.1	1100

Table 1. Microwave heating of some oxides and sulfide compounds

Further, Wong (1975) [18]and Tinga (1988[19], 1989[20]) reported the microwave heating behavior of several metal oxides. These results were compared with published data; and classified based on heating rate into hyperactive, active, difficult-to-heat and inactive. Table 2 represents the compilation results. They demonstrated that microwave energy could be effective in the heating of minerals and inorganic compounds.

Material classification		Heating rate reported (°C/min)	Max. Temp. (°C)
Hyperactive Materials	UO_2	200 (°C/s)	1100
	MoS_2	150 (°C/s)	900
	C(charcoal)	100 (°C/s)	1000
	Fe_3O_4	20 (°C/s)	500-1000
	FeS2	20 (°C/s)	500
	CuCl	20 (°C/s)	450

Material classification		Heating rate reported (°C/min)	Max. Temp. (°C)
Active	Ni_2O_3	400	1300
	CoO_3	300	900
	CuO	200	800
	Fe_2O_3	170	1000
	FeS	135	800
	CuS	120	600
Difficult to Heat	Al_2O_3	80	1900
	PbO	70	900
	MgO	33	1300
	ZnO	25	1100
	MoO_3	15	750
Inactive	CaO	5	200
	$CaCO_3$	5	130
	SiO_2	2 – 5	70

Table 2. Classification of some reagent grade materials based on microwave heating rate

Perhaps the most important of the early work was that of Chen et al., 1984,[21] who investigated the reaction of 40 minerals to microwave exposure in a waveguide applicator, which allowed the mineral samples to be inserted in an area of known high electric field strength. Though by this time, it was already known that microwaves would heat some minerals selectively; this work further showed that microwave heating is dependent on the composition of the mineral, and thus elemental substitutions would affect the behavior of a mineral in an electric field. An example of this was noted with sphalerite, where high iron sphalerite would eventually heat quite well after a period of slow heating at low temperatures, but that low iron sphalerite did not heat readily. From the large number of minerals tested, it was noted that most silicates, carbonates and sulfates, and some oxides and sulfides are transparent to microwave energy, while most sulfides, arsenides, sulfosalts and sulfarsenides, and some oxides, heat well when subjected to microwave irradiation.

More recently, the US Bureau of Mines reported test results of microwave heating a number of minerals and reagent grade inorganic compounds with 2450 MHz (McGill and Walkiewicz, 1987[22], Walkiewicz et al., 1988[23]). The test results revealed that the highest temperatures were obtained with carbon and most of the metals oxides: NiO, MnO_2, Fe_3O_4, Co_2O_3, CuO and WO_3. Most metal sulphides heated well but without any consistent pattern. Metal powder and some heavy metal halides also heated well; gangue minerals such as quartz, calcite and feldspar did not heat. This study also revealed that rapid heating of ore minerals in a microwave transparent matrix generated thermal stress of sufficient magnitude to create micro-cracks along mineral boundaries. This kind of micro-cracking has the potential to improve grinding efficiency as well as leaching efficiency.

Chunpeng et al. (1990)[24] conducted microwave heating tests on several oxide, sulfide and carbonate minerals. All tests were conducted on a 50.0 g powder (-200 mesh.) sample per batch with an input microwave power of 500 W of 2450 MHz frequency and constant exposure time (4 min). Test results are shown in Table 3. These results indicate that the majority of oxide and sulphide minerals heated well.

Minerals	Chemical composition	Temperature (°C)
Jamesoite	$Pb_2Sb_2S_5ZnS$	>850
Titanomagnetite	$xTiO_2.\ yFe_3O_4$	>1000
Galena	PbS	>650
Chalcopyrite	$CuFeS_2$	>400
Pentlantite	$(FeNi)_{(1-x)}S_8$	>440
Nickel pyrrhotite	$(FeNi)_{(1-x)}S$	>800
Cu–Co sulphide Concentrate	$xCu_2S.\ yCoS$	>800
Sphalerite	ZnS	>160
Molybdenite	MoS_2	>510
Stibnite	Sb_2S_3	Room temp
Pyrrhotite	$Fe_{(1-x)}S$	>380
Bornite	Cu_3FeS_4	>700
Hematite	Fe_2O_3	>980
Magnetite	Fe_3O_4	>700
Limonite	$mFeO_2.nH_2O$	>130
Cassiterite	SnO_2	>900
Cobalt hydrate	$CoO.nH_2O$	>800
Lead molybdenate	$PbMoO_4$	>150
Iron titanite	$FeTiO_3$	>1030
Rutile	TiO_2	Room temp
Lead carbonate	$PbCO_3$	>180
Zinespar	$ZnCO_3$	>48
Siderite	$FeCO_3$	>160
Serpentine	$Mg(Si_4O_{10})(OH)_3$	>200
Melaconite	$Cu_2Al_3(H_{(2-x)}Si_2O_3)(OH)_4$	>150
Antimony oxide	Sb_2O_3	>150

Table 3. Effect of microwave heating on the temperature of various minerals (500 W, 2450 MHz, 4min radiation)

Interaction of microwave with minerals is poorly understood. Thus, a fundamental understanding of how microwave energy interacts with minerals is the key to unlocking the technology for use in mineral processing industries. To shed more light on the subject of the interaction of microwave with minerals, Barani et al., 2012,[25] studied the effect of sample factors, such as volume, surface area, size and shape, aspect ratio on the magnitude and uniformity of power absorption by iron ore and water samples and compared obtained data. The results showed that for water heating, with increasing in sample volume from 200 to 1000 cm^3 the microwave energy absorbed by water was increased from 71.27 to 100%, also with increasing in sample surface area from 50.24 to 78.50 cm^2 the microwave, energy absorbed by water was increased from 76.36 to 89.09%. With increasing iron ore sample volume without increasing in surface area, the microwave absorption was constant whereas with increasing in sample surface area from 50.24 to 126.6 cm^2, the microwave energy absorbed by iron ore was increased from 36.6 to 61.82%. The maximum temperature for iron ore material was occurred at 5.7 cm distance from the center whereas the maximum temperature for water sample was occurred at 5cm distance from the center.

2.2. Microwave assisted ore grinding

Walkiewicz et al. (1988[23], 1991[26]) demonstrated that the rapid heating of ore containing microwave energy absorbing minerals in a non-absorbing gangue matrix generated thermal stress. This thermal stress caused micro fracturing along the mineral grain boundaries; as a result, such an ore sample becomes more amendable to grinding. According to these authors, the grinding operation (comminution) consumes 50%–70% of energy used in mineral processing operations. Again, the energy efficiency of a conventional grinding operation is approximately 1%. They demonstrated that microwave preheating of an iron ore improved grinding efficiency by 9.9% to 23.9%. However, this improvement was not enough to compensate for the energy consumption of the microwave preheating.

Walkiewicz et al., 1993,[27] investigating the effect of power level on Bond work index, found that the larger temperature gradients associated with the more rapid development of heat within the particle grains because of higher microwave powers, led to a larger decrease in ore strength than for exposure to lower microwave powers.

Tavares and King, March 1996,[28] investigating samples of specific iron, taconite and titanium ores in a multimode cavity using a low power input of between 0 and 1.2 kW, compared the strengths of untreated ore with that of ores treated both conventionally and with microwaves. It was observed that in all cases the thermal treatments affected the ore favourably in terms of both reductions in fracture energy and increased damage, however, there was very little difference between the results for the conventional and microwave treatments, with the exception of a greater reduction in fracture energy of the iron ore and greater damage to the titanium ore from microwave treatment. From examinations of the single particle breakage functions, it was further seen that the thermal pretreatments resulted in a shift in the top of the breakage function to smaller sizes without an increase in the production of very fine material, and also that the microwave treated ores tended to produce a greater shift in the top of the breakage function than conventionally treated ores. It was concluded that this change in fragmentation pattern, together with observations from image analysis of a 50% increase in grain boundary fracture in the microwaved iron ore, might result in improved liberation. Later tests by the same authors (Tavares and King, August 1996)[29] on a copper ore showed no difference between the fracture energies of microwave pretreated and untreated material, though it was noted that there was a slight indication of grain boundary fracture around the sulfide grains. It is not stated what kind of microwave treatment was used, however, and thus these results are not comparable to those of other workers.

Work on the grind ability of coal by Marland et al., 2000[30], indicated that reductions in work index of up to 50% occur after microwave pretreatment. The greatest strength reductions were obtained from lower ranked coals, and it was suggested that this was most likely due to the higher inherent moisture content of such coals, with gaseous evolutions of water and volatile matter the main causes of damage to the coal particles. It was also found that microwave radiation affected the calorific value to the same extent as would be expected from conventional drying procedure, and it was concluded that the application of microwave treatment did not alter the fuel potential of coal.

Kingman et al., 2000[31], encompassing tests on several commercially exploited ores to investigate the influence of ore mineralogy on microwave assisted grinding showed that the most responsive ores were those with a consistent mineralogy, containing good absorbers in a transparent gangue, while those with small lossy particles that are finely disseminated in discrete elements were shown to have the worst response in terms of reduction in required grinding energy. One extremely important result from this paper was the suggestion that purpose built microwave cavities may be important in making the treatment of ores more economically viable.

Wang and Forssberg, 2000,[32] performed tests on three ores (i.e. limestone, dolomite and quartz) to investigate their microwave heating behavior and subsequent grindability during dry ball milling, after pretreatment. Each ore was crushed and sized into three fractions for testing, these being -9.75+5.75 mm, -4.7+1.6 mm and -1.6 mm. It was noted that the particle size of the material undergoing thermal pretreatment had a significant effect on the heating behavior and subsequent grindability of two of the ores, with tests on the quartz and limestone material showing that the microwave pretreatment was only effective for the -9.5+4.75 mm material, which then subsequently showed improved grindability. Below 4.75 mm, little or no effect was seen, and it was suggested that this was due to conductive heat transfer which plays a more important role in heat loss from smaller particles. It was also found that increasing the exposure time led to a further increase in the grindability of these two ores. Dolomite showed little reaction to microwave pretreatment during subsequent dry milling experiments. Tests were also performed to determine the degree of liberation of sulfide minerals in a low grade copper ore (0.22-0.4% Cu) from Aitik after crushing. SEM photomicrographs showed that thermal stress cracks occurred readily along the sulfide-gangue mineral grain boundaries, and image analysis software showed a substantial increase in the liberation of sulfide minerals in the ore matrix with microwave pretreatment prior to crushing.

Vorster et al, 2001[33], performed several tests on a massive copper ore and a massive copper-zinc ore, both from Neves Corvo in southern Portugal, using a 2.6 kW multimode cavity operating at 2.45 GHz. Quenching after 90 seconds of microwave exposure led to a 70% reduction in the work index of the massive copper ore. The effect of quenching was also illustrated with tests on the massive copper-zinc ore, where after 90 seconds of microwave exposure with no quenching, a reduction of 50% in the strength of the ore was obtained, while the addition of quenching directly after microwave treatment led to a further 15% reduction in work index. Copper flotation trials showed that no benefit in terms of improved copper recovery was seen after microwave treatment, and it was concluded that the improved liberation after microwave treatment which was noted from SEM analysis, was most likely offset by some surface oxidation of the recoverable sulfide minerals.

Kingman et al., 2004[34], investigated the treatment of a copper carbonatite ore from a mine in South Africa using a single mode, high power applicator (i.e. a variable power input of up to 15 kW). Their results showed that a sort of threshold value existed for the power input into the system, which once passed, caused serious damage to the particle in a very short treatment time (< 0.5 seconds). The importance of this discovery is best seen when the values are turned into values of power densities within the valuable minerals, in which case these values are no longer specific to a certain microwave system, allowing the design of any system with the goal of obtaining these power densities. It was shown that reductions of up to 30% in grinding energy could be achieved with microwave energy inputs of less than 1 kWh/t. QEMSCAN analysis of the product of drop weight tests also showed a decrease in the amount of locked and middling copper sulfides in the +500 μm size classes.

Amankwah et al., 2005[35], performed tests on samples of a gold ore containing quartz, silicates and iron oxides with a head grade of 6.4 g/t of gold, using 2 kW of power in a multimode cavity. It was seen that the microwave treatment resulted in a maximum reduction of 31.2% in crushing strength and a reduction of 18.5% in work index. SEM analysis clearly showed that microwave induced fractures were occurring in the ore, and an improvement of 12% in gold recovery from gravity separation tests showed that this resulted in the liberation of the gold at coarser particles sizes during comminution.

Scott et al., 2007, studied the effects of microwave treatment on the liberation spectrum of a rod-milled South African carbonatite ore.The treated ore was processed for 0.5 s at 10.5 kW in a single mode microwave cavity in batches of 1 kg. The treated and untreated ore were subsequently grinded to 80%–800 μm. The microwave treated ore showed a significant increase in the amount of liberated copper minerals in the relatively coarse particle size range (106 to 300 μm). Similar significant shifts in the liberation spectra were noted for all the minerals in the ore. It is inferred that microwave treatment induces changes in the fracture pattern, favoring liberation of microwave susceptible minerals at larger particle sizes.

Koleini et al., 2008[36], investigated the effect of microwave radiation on the comminution of an iron ore. Iron ore material was preheated for different time at 1000W in a multi-mode microwave oven. Comparative bond rod mill work index was used to determine the effect of this process on the grinding energy required for size reduction of the material in a laboratory rod mill. It is shown after 1, 3 and 5 minute radiation respectively, the amount of 12, 34 and 46% reduction in work index was achieved. Microwave exposure followed by water quenching is shown after 1, 3 and 5 minutes radiation respectively, the amount of 19, 38 and 50% reduction in work index was achieved.

Barani et al., 2010[37], studied the influence of microwave pre-treatment on iron ore breakage. Drop weight tests were used to quantify the change in strength in terms of reduction in required comminution energy. The drop weight test parameters of untreated iron ore was compared to microwave-treated iron ore under the same experimental conditions and it was found that microwave-treated materials is softer than untreated in terms of the impact breakage parameter values and the abrasion breakage parameter values. After microwave treatment, about 100% increases in abrasion breakage parameter was achieved while maximum increasing in impact breakage parameter was 36%. It seems that microwave treatment is more effective for abrasion breakage mechanism; because abrasion is, a surface phenomenon and microwave treatment is more effective at surface heating

Koleini et al, 2012[38], studied the effect of microwave treatment on the grinding kinetics of an iron, using mono-sized materials of −2.360+2.0 mm, −1.400+1.180 mm , −1.0+0.850 mm and −0.355+0.300 mm. Microwave-treated samples were treated in a multimode microwave oven with 1100 W input power. The grinding tests were conducted using a laboratory ball mill under identical conditions to allow a comparative analysis of the results. The specific rates of breakage (Si) and cumulative breakage distribution function (Bi,j) values, as grinding breakage parameters, were determined for those size fractions of untreated and microwave-treated feeds. It was determined that breakage of iron ore followed a first-order behavior for fine feed sizes and deviated from first order for coarse feed size. The specific rate of breakage parameters of untreated iron ore was compared with microwave-treated iron ore under the same experimental conditions and it was found that microwave-treated materials break faster than untreated in terms of the Si and A values. Breakage parameters showed that treated materials produce more coarse material than untreated material in terms of the γ value of Bi,j.

2.3. Breakage mechanism

Walkiewicz et al., 1988[23], showed that thermal stress fracturing along grain boundaries was induced in some samples after microwave heating, and suggested that this could significantly influence not only the grindability of microwave treated ores, but mineral liberation as well.

Work by Tinga, 1988[19], in the field of microwave sintering suggested that preferential heating of grain boundaries occurs. This should be the case for any high loss dielectric grain of reasonable diameter embedded in a relatively low loss host material. Effects such as conduction losses and the rate of heating do play a role, however, and care should be taken before assuming this is true for any particular situation. Tinga, 1988[19], also stated that the single most important factor when considering microwave heating was the design of the applicator, where choosing the wrong applicator for a task will mostly likely result in very few of the expected benefits of microwave processing being seen, and therefore very little improvement in results from the treatment versus those of conventional practices.

Salsman et al., 1996[39], used a finite element numerical model of a single pyrite particle in a calcite matrix to further investigated the phenomenon of thermally assisted liberation using microwave energy. Using power densities, which are likely to be possible within the pyrite grains, it was seen that large tensile stresses, exceeding the tensile strengths of most common rock material, were generated along the pyrite-calcite interface. It was discovered that a decrease in either particle size or in the grain size of the microwave susceptible mineral inclusions, led to a decrease in the intergranular stresses developed within the particles. The influence of power density on the absorption of microwave energy by minerals was also investigated, and it was found that by using short concentrated microwave pulses to increase the power density within the material, substantially higher stresses could be generated within the particles at the same power inputs.

Whittles et al., 2003[40], investigated the effect of power density on the microwave treatment of ores, using finite difference techniques to model microwave heating, thermal conduction, thermal expansion, thermally induced fracturing and strain softening of a particle containing dispersion of 2 mm square pyrite grains in a 15 mm by 30 mm calcite host matrix. Simulations were also performed to determine any change in the uniaxial compressive strength of the particle after microwave heating. It was shown that power density is an important factor in microwave treatment of ores, with the application of high power densities resulting in much greater damage to the particle. It was concluded that utilizing high power densities for shorter times could also drastically reduce the microwave treatment energy required to below 1 kWh/t.

Jones et al., 2005[41], also investigated the effect of microwave treatment through numerical simulations of a system of microwave absorbing pyrite grains in a microwave transparent calcite host. An important result of this work was the verification and explanation of the observations of Wills et al., 1987[10], who determined that regularly shaped mineral inclusions with smooth boundaries were much more likely to result in thermally induced intergranular fracture than irregular grains which tend to be damaged by transgranular fracturing as a direct result of thermal treatment. It was determined that for spherical absorbing grains the occurrence of transgranular fracture is highly unlikely as the symmetry of the grain ensures that the compressive stresses generated inside the microwave absorber are equal in all directions, thus reducing the likelihood of shear stresses developing within the grain. As grain shape deviates from spherical, the likelihood of transgranular fracture rises. It was also seen that as the grain size of the microwave absorber decreased, conduction losses resulted in lower temperatures being reached within the absorbing grain at the end of the same exposure time. This resulted in lower stresses being generated around the absorbing grain, with less damage to the host particle as a result.

2.4. Microwave treatment and magnetic properties of minerals

Florek et al., 1996[42], carried out a study of the effect of microwave treatment on the magnetization of iron ore minerals. It was concluded that the surface characteristics and magnetization of iron ore minerals alter after microwave radiation.

The effect of microwave radiation upon the mineralogy and magnetic processing of amassive Norwegian ilmenite ore was per- formed by Kingman et al. 1999[43]. It has been shown that short periods of exposure can cause fracture at grain boundaries, which leads to the formation of inter-granular fractures. This fracture coupled with an increase in remnant magnetization of the ilmenite mineral has been demonstrated to give rise to an increase in both concentrate grade and valuable mineral recovery. However, the study has also indicated that process efficiency can be effected with over exposure to microwave radiation.

Kingman and Rowson, 2000[44], showed that a number of minerals, e.g. chalcopyrite, hematite and wolframite, not only heat readily during exposure to microwaves, but also exhibit a considerable increase in the magnetic susceptibility after being exposed to 650W microwave radiation.

Cui et al., 2002[45], carried out an investigation to study the changes in magnetic properties after roasting to the different types of minerals contained in the oil sands tailings. It was observed that the magnetic susceptibility of ilmenite increased after either oxidation or reduction roasting. For hematite, reduction roasting increased its magnetic susceptibility and oxidation roasting did not seem to have any effect.

Sahyoun et al., 2003[46], investigated the influence of conventional heat treatment and microwave radiation on chalcopyrite. There was a significant increase in the proportion of material recovered to magnetic fraction and magnetic susceptibility with conventional heating time. XRD analysis detected phase changes in conventional heat-treated chalcopyrite, which increases the magnetic susceptibility of the ore and enables its effective magnetic separation, which is impossible to achieve in its original state. With microwave treatment, the magnetic susceptibility increases and the proportion of material recovered to the magnetic fraction on the induced rolls is also increased. However, XRD analysis failed to detect any phase changes. A possible explanation for this observed behavior can be drawn that the more magnetic component has been formed by microwave treatment is below the threshold of detection of the XRD analyzer.

Uslu et al., 2003[47], investigated the effect of microwave heating on magnetic processing of pyrite. The microwave treated pyrite samples of −0.420mm fraction were subjected to magnetic separation at magnetic field intensities of 0.1, 0.3 and 0.5T. It was found that pyrite was converted to such ferromagnetic minerals as pyrothite and γ-hematite, and magnetic separation recovery was improved after microwave treatment.

Znamenackova and Lovas., 2005[48], showed that after 10 min pre- treatment of weakly paramagnetic ore in a microwave oven with maximum power of 900 W, essential change in the magnetic properties of the ore samples occurred and after 15 min, a rapid increase of magnetic susceptibility value was observed, showing the intensive decomposition of siderite. Finally, after 40 min of heating, a microwave sintering of powder grains in the form of agglomerates with molten mass was observed.

Waters et al., 2007[49], investigated the effect of microwave radiation on the magnetic properties of pyrite. After treatment with a conventional multimodal reactor (2.45GHz and

1900W) for 120 s, the recovery of pyrite in the magnetic fraction after separation increased from 8% (wet) and 25% (dry) to greater than 80% for both process streams. The improvement in the magnetization of the sample has also been noted, determined using a vibrating sam- ple magnetometer (VSM). After exposure to microwave radiation, the magnetization of the mineral sample was increased.

Barani et al., 2011[50], investigated the effect of microwave radiation on the magnetic properties of an iron ore. Four Iron ore samples were used in this research. Three samples were treated for 30, 60 and 120 S respectively, in a multi-mode microwave oven with a frequency of 2.45GHz and a maximum power of 1100 W. The magnetizations of non-treated and microwave-treated samples were determined using a vibrating sample magnetometer (VSM). With increasing in radiation time to 60 S, the total magnetism saturation and remnant magnetization of the samples were increased. The results show that with further increasing in microwave radiation time up to 120 S, localized sample melting was occurred and the total magnetism saturation and remnant magnetization were decreased. The results showed that the sample composes ferromagnetic and paramagnetic fractions. With increasing in microwave radiation time the magnetic susceptibility of the paramagnetic fraction was decreased from 0.0111 to zero whereas the magnetic susceptibility of the ferromagnetic fraction initially was increased from 0.0687 to 0.3879 then decreased to 0.1894 (at 120 S radiation time). It was confirmed that microwave radiation has a significant effect upon magnetic properties of iron ore. However, there is a limited condition, excessive radiation has a negative effect and reduces the magnetic susceptibility of iron ore.

2.5. Microwave assisted pretreatment of refractory gold concentrate

Gold is considered to be refractory when it cannot be easily recovered by alkaline cyanide leaching. The vast majority of refractory gold occurs in sulphidic minerals such as pyrite (FeS_2), arsenopyrite (FeAsS) and pyrrhotite (FeS). Generally, refractory gold concentrate or ore is pretreated by roasting, O_2-pressure leaching or bacterial leaching, to render it amenable to gold recovery by alkaline cyanide leaching (Haque, 1987a, b)[51, 52]. Because microwaves in general heat sulphidic minerals easily, it should be possible to pretreat sulphidic refractory gold concentrate by microwave energy. (Haque 1987a, b)[51, 52] conducted laboratory-scale microwave pretreatment tests in air on a typical arsenopyritic refractory gold concentrate. More than 80% of As and S were volatilized as As_2O_3 and SO_2, whereas iron was oxidized into hematite (Fe_2O_3) at 550°C (uncorrected). Alkaline cyanide leaching of the calcine yielded 98% Au and 60% Ag extractions.

To avoid the formation of As_2O_3 and SO_2 this author conducted microwave calcination tests on the concentrate in a nitrogen atmosphere in a sealed silica tube. The major products were FeS, arsenious sulphide (As_2S_3) and sulphur (S). In addition, this author conducted microwave heating tests on a mixture of this concentrate and (NaOH), No SO_2 and As_2O_3 evolved during microwave heating of this mixture; instead water soluble products such as; Na_3AsO_4 , Na_2SO_4, $FeSO_4$ were formed. The microwaved solids were leached with water at 75°C. After phase separation, the residue was leached with alkaline cyanide solution, and yielded 99% Au and 79% Ag extractions. These results opened up a wide range of possibilities for investigation (Woodcock et al., 1989[53]). Non typical refractory gold ore, such as carbonaceous gold ore, some goethite bearing gold tailings, etc., have also been successfully pretreated by microwave heating (Author's unpublished results). Currently, EMR Microwave Technology, Fredericton, N.B., Canada is conducting pilot-scale microwave pretreatment tests on various kinds of refractory gold concentrates, ores and tailing.

Al-Harahsheh et al., 2005[54], have investigated the leaching kinetics of chalcopyrite under the influence of microwave treatment. Comparison of the amount of copper recovered from chalcopyrite under conventional and microwave heat treatment show marginal, but consistent, improvements in copper recovery when using microwave treatment as opposed to conventional treatment. It was suggested that the increase in copper recovery with microwave leaching was due to localized higher temperatures around the outer shell of the leaching solution as a result of the high dielectric loss factor (and thus low penetration depth) of the solution, and also selective heating of the outer skin of the chalcopyrite particles due to the high conductivity of this material.

Amankwah et al., 2005[55], performed tests on samples of a gold ore containing quartz, silicates and iron oxides with a head grade of 6.4 g/t of gold, using 2 kW of power in a multimode cavity. It was seen that the microwave treatment resulted in a maximum reduction of 31.2% in crushing strength and a reduction of 18.5% in work index. SEM analysis clearly showed that microwave induced fractures were occurring in the ore, and an improvement of 12% in gold recovery from gravity separation tests showed that this resulted in the liberation of the gold at coarser particles sizes during comminution.

Nanthakumar et al., 2007[56], investigated microwave roasting of a double refractory gold ore as an alternative method and the results were compared to those obtained by conventional roasting. The compositional changes of the ore during roasting were determined by thermo gravimetric analysis (TGA). In addition, both the real and the imaginary permittivities, which determine the amount of energy absorbed by the ore and the heating rate of the ore respectively, were evaluated. In addition, the microwave heating behavior was studied. Conventional and both direct and indirect microwave roasting tests were performed and in all the cases, the pyrite was readily converted into hematite. Direct microwave roasting could not remove the organic carbon. Indirect microwave roasting was conducted using magnetite as a susceptor and preg-robbing was eliminated when about 94% of the organic carbon was removed. For both conventional and indirect microwave roasting, gold recoveries of over about 98% were achieved after cyanide leaching. For microwave roasting, both the total carbon removal rates and the heating rates were higher and the specific energy consumptions were lower than the corresponding values for conventional roasting.

Amankwah et al., 2008[57], studied microwave roasting of a double refractory flotation concentrate to oxidize both the sulfides and the carbonaceous matter. The concentrate was characterized by thermo gravimetric and infrared analysis and the microwave absorption characteristics were quantified by determining the permittivities. The microwave heating behavior studies showed that the sample temperature increased with increasing incident microwave power, processing time and sample mass. Due to the hyperactive response of the concentrate to the microwaves, a low incident power of 600 W was found to be suitable for roasting, as higher powers resulted in sintering and melting of the concentrate. The gold extraction values after cyanidation were over 96% and these were similar to those obtained by conventional roasting. The main advantages of microwave roasting were that both the total carbon removal rates and the heating rates were higher and the specific energy consumptions were lower.

Ma et al., 2008[58], investigated removal of sulfur and arsenic from refractory flotation gold concentrates, bearing with 14.95% of As and 27.85% of S, by microwave roasting. Cooling patterns of the roasted products obviously affected the removal effects under oxygen-free roasting atmosphere. The highest removal occurred by crucible-uncapped cooling pattern, followed by the so-called half-open cooling pattern, and the crucible-capped cooling pattern attained the lowest removal. The mid pattern would be preferred because it could avoid spontaneous ignition compared to the crucible-uncapped cooling one. Roasting temperature showed obvious effect only above 450°C, increasing with the roasting temperature. However, desulfur was much more difficult than de-arsenic. Under oxygen-free roasting atmosphere, 95% of arsenic was removed when roasted for 40 minutes at 550°C, while the desulfur rate was only about 40%. Comparatively, the removal of sulfur dramatically reached above 90% in oxidizing atmosphere. Additionally the roasted products were analyzed by XRD.

2.6. Dielectric properties of minerals in microwave fields

The nickeliferous laterite ores, in which the nickel occurs in oxide form, represent a significant potential resource of metallic nickel. However, in comparison to the nickel-containing sulfide ores, the extraction costs are relatively high and thus it will be necessary to develop new processing techniques, which are both technically and economically viable. Pickles et al., 2004[59], investigated the potential application of microwaves for the heating of a nickeliferous limonitic laterite ore ($(Fe,Ni)O(OH).nH_2O$) was investigated. Firstly, since the nickeliferous limonitic laterite ore contains considerable moisture, both free and combined, then thermogravimetric analysis (TGA) was performed in order to characterize the changes, which result from the dehydration processes. Derivative thermogravimetric analysis (DTGA) curves were calculated from the TGA data. Secondly, the real (ε') and imaginary (ε'') permittivities of the ore were measured at frequencies of 912 and 2460 MHz at temperatures up to about 1000 °C using the cavity perturbation technique and these results were related to the DTGA curves. Also, the loss tangent ($\tan\delta = \frac{\varepsilon''}{\varepsilon'}$) was calculated from the permittivity data. Finally, the microwave heating behaviour of the nickeliferous limonitic laterite ore was determined at 2460 MHz. The results show that the both the real (ε') and imaginary permittivities (ε'') and the loss tangent ($\tan\delta = \frac{\varepsilon''}{\varepsilon'}$) increase with temperature and change as both the free and the combined moisture are removed. The permittivities (ε' and ε'') increased with increasing slope of the TGA curve and vice versa during the goethite to hematite dehydroxylation reaction, where there was a maximum in the permittivities (ε' and ε''). It is proposed that these changes, which occur during the dehydroxylation reaction, are a result of the liberation of hydroxyl units from the goethite structure.

Cumbane et al., 2008[60], has been used a measurement system, comprising a circular cylindrical TM_{0n0} cavity and based on a perturbation technique, for the determination of dielectric properties of five powdered sulphide minerals, which were measured at frequencies of 615MHz, 1410 MHz and 2210 MHz. The complex permittivity was measured from ambient temperature to 650 °C. The dielectric properties of galena and sphalerite exhibit little variation with temperature up to 500 °C. The dielectric properties of pyrite, chalcocite and chalcopyrite, show significant variation with temperature. These are related to composition and phase transformations during heating and were demonstrated by thermo-gravimetric analysis.

2.7. Microwave assisted carbothermic reduction of metal oxide

The vast majority of heavy metals oxides and carbon, as charcoal or coke, respond to microwave heating. Therefore, the microwave assisted carbothermic reduction of metal-oxides is possible. If the metal oxide is low lousy (i.e., poor receptor to microwave energy) then added carbon plays the role of microwave heating accelerator. Various researchers have demonstrated that iron oxides (hematite Fe_2O_3, magnetite Fe_3O_4) mixed with carbon (charcoal or coke) could be reduced to metallic iron (Standish and Worner, 1991[61], Gomez and Aguilar, 1995[62]).

To compare conventional and microwave reduction, Standish et al. (1990, 1991)[63, 61] conducted reduction tests on identical sample mixtures of hematite ore fines, coke and lime powder. A sample of each mixture was heated in an electrically heated muffle furnace at 1000°C, and in a 2450 MHz microwave oven at a power of 1.3 kW. The sample temperature was measured with a thermocouple inserted in the sample and the test was terminated when the temperature reached 1000°C. The results showed the microwave heating rate was much higher than the conventional heating rate. Some phase changes were observed and these might have enhanced the heating rate. Standish et al. (1991, 1990)[61, 63] concluded based on rational assumptions for capital and operating costs, that a microwave reduction process could save 15% to 50% over a conventional operation. Chunpeng et al., 1990[64], also conducted microwave assisted carbothermic reduction on titanomagnetite concentrate. A powdered sample of titanomagnetite concentrate mixed with lignite powder and $CaCO_3$ was heated by microwave power of 500 W at 2450 3 MHz. These results, compared with those generated by conventional heating test, confirmed that the reduction rate of metal oxide by microwave heating was faster than by conventional heating. Beside the carbothermic reduction of iron oxides the researchers used microwaves to smelt rare earth magnet alloys, a high value product difficult to produce by conventional techniques. Although these alloys could be produced in a microwave furnace, the furnace needed design changes to eliminate the formation of gas plasma over the melt. Moreover, a suitable microwave transparent material was needed to contain the smelt at high temperature.

2.8. Microwave assisted drying and anhydration

Materials and products such as agricultural, chemical and food product, textile, paper, lumber and many more (Cook, 1986[65], Schiffmann, 1987[66], Doelling et al., 1992[67]).Generally, drying refers to the removal of physically adsorbed solvent such as water, acid or high vapour pressure organic substance (e.g., alcohol, acetone, ether, halogenated hydrocarbons, aromatics, etc.). Anhydration refers to the removal of water chemically bound to a substance present intermolecularly as well as to the intramolecular elimination of water from hydroxy or carboxylic compounds. It was observed that the dielectric loss factor of a material to be dried often decreases with the loss of solvent (Schiffmann, 1987[68]). Unpublished results indicate that microwave heating can remove water from hydrated magnesium chloride ($MgCl_2.7H_2O$) and convert goethite (O=Fe–OH) into hematite (Fe_2O_3), (Haque, 1998[68]).

If both the solvent and the substance to be dried are transparent to microwaves (i.e., no heating by microwave energy), then a suitable microwave heat accelerator, such as carbon, magnetite or silicon carbide must be added to the system to heat the added material to volatilize the solvent. This heating concept may be applied in the removal of volatile contaminants from soil, or even ore material (George et al., 1994[69]).

2.9. Microwave assisted minerals leaching

Analytical chemists have used microwave heating devices routinely for the dissolution of metals, minerals and various chemical products in chemical analysis (Matthes et al., 1983[70], Kingston and Jassie, 1985[71]). As mentioned earlier, microwave heating is material specific, offers a faster heating rate and consequently a faster dissolution rate than conventional heating. In fact, the principle of the dissolution of analytical samples has been applied to the leaching of various minerals contained in an ore or concentrate sample. Kruesi and Frahm, 1982[72] and Kruesi,1986[73] conducted microwave assisted leaching of lateritic ores containing oxides of nickel, cobalt, and iron. The metals of these mineral components were converted into their chlorides by microwave heating (1200 W, 2450 MHz, N_2 atmosphere) a mixture of the ore and ammonium chloride between 177°C and 312°C for 4–5 min, followed by water leaching at 80°C for 30 min. Nickel and cobalt extractions were 70% and 85%, respectively, and are comparable with roasting at 300°C in a conventional rotary kiln for 2 h. Similarly, copper ores or concentrates containing oxidic and/or sulphidic minerals were solubilized by microwave heating a mixture of the ore or concentrate and ferric or ferrous chloride between 350°C and 700°C, followed by hot brine leaching. Copper extraction was 96% (Kruesi and Frahm, 1982[74]). To study nickel extraction, Chunpeng et al., 1990[24], conducted dry way chloridization of pentlandite concentrate with ferric chloride by microwave heating (500 W, 2450 MHz) in a chlorine atmosphere for 8–23 min, followed by aqueous leaching at pH 2 for 30 min. The maximum nickel recovery (~99%) was obtained from the sample heated for 14–17 min.

Peng and Liu, 1992a[74], 1992b[75], applied microwave energy in the leaching of sphalerite with acidic ferric chloride ($FeCl_3$ –HCl). Various leach parameters; such as 3 temperature, particle size and ferric chloride concentration were studied. Test results demonstrated that the leaching rate of zinc increased with temperature in both microwave and conventional heating systems. They reported 90% zinc extraction when the leach conditions were at 0.1 M HCl, 1.0 M $FeCl_3$, 60 min microwave heating at 95°C. Under similar conditions conventional leaching yielded only 50% zinc extraction. Weian, 1997[74], conducted microwave assisted acidic ferric chloride leaching of a copper sulphide concentrate. The principal copper minerals in this concentrate were chalcocite (Cu_2 S) and chalcopyrite ($CuFeS_2$). The leach slurry was heated directly by microwaves (700 W, 2450 MHz) for various lengths of time. Copper recovery reached 99% after 40–45 min of microwave heating whereas conventional heating required 2h to reach the same level of extraction. This author concluded that microwave assisted leaching provided a faster dissolution rate of copper and overcame the detrimental effect from elemental sulphur build-up on the mineral surface during the leaching of the copper sulphide concentrate.

In the recovery of copper from a chalcopyritic concentrate (30.1–30.3% copper) Antonucci and Correa,1995[75], conducted a sulfation reaction by microwave heating (2450 MHz for laboratory tests and 915 MHz for semi pilot scale tests) a paste-like mixture of the concentrate and sulphuric acid followed by water leaching at 60°C and pH 1.6. Semi pilot scale tests were conducted in a 35 L capacity Teflon-lined cylindrical rotatory reactor. The whole setup was placed in a multimode applicator and microwaves (915 MHz) were applied. There was an opening on top of the reactor through which the charge (conc. + H_2 SO4) was fed and which also served as the outlet for the recovery of gas and elemental sulphur. The test results indicated that higher copper extraction could be achieved at higher sulphuric acid dosage. These authors concluded that copper extraction>96% could be achieved by microwaving a mixture containing 1.80 kg acid/kg conc. The process gave elemental sulphur and cupric sulphate. Antonucci and Correa, 1995[79], also commented that although this process demanded more energy than the conventional smelting process the production of elemental sulphur was advantageous.

2.10. Microwave assisted spent carbon regeneration

Currently, more and more gold ore processing industries are using activated carbon in CIP (carbon in pulp) or CIL (carbon in leach) operation. The carbon is regenerated after each cycle of adsorption and desorption of gold cyanocomplex. Usually, this spent carbon is regenerated by washing with a mineral acid followed by heating at high temperature (600°C to 750°C) in an externally heated rotary kiln (Avraamides et al., 1987[76]).

Haque et al., (1991[77], 1993[78]), conducted laboratory scale carbon regeneration tests by microwave (2450 MHz) heating and confirmed the feasibility of spent carbon regeneration by microwave heating. Subsequent pilot scale carbon regeneration tests data (915 MHz) demonstrated that microwave regenerated carbon performed well or better than conventionally regenerated carbon (Bradshaw et al., 1997[77]). Currently, Ontario Hydro, Toronto, Ontario, Canada is marketing this technology.

2.11. Microwave assisted waste management

Processing industries invariably generate waste material; mine-milling industries are no exception. To mitigate the danger presented by the constituents of the waste technologies are being investigated to minimize the waste generated and to provide safe handling, transportation, storage, destruction, removal or disposal of the hazardous waste. Currently, microwave energy is showing considerable potential in the management of a vast array of gaseous, liquid and solid wastes (Wicks et al., 1995[78]). Mine milling operations generate large volumes of solid waste with acid generation potential, liquid waste containing acid, toxic heavy metals and non-metals, cyanide, ammonia, organics etc., and gaseous wastes such as, sulphur dioxide (SO_2), hydrogen sulphide (H_2S), ammonia (NH_3), oxides of nitrogen (NO_x).

Cha (1993) demonstrated in laboratory scale tests that SO_2 and NO_x in industrial off-gas can be decomposed into·elemental nitrogen and sulphur, and a mixture of carbon dioxide and carbon monoxide. The first step of the process involves passing the off-gas stream containing SO_2 and/or NO_x through a column packed with activated carbon to adsorb the toxic gases. The loaded carbon column is then heated by microwaves and the resulting CO, CO_2 and N_2 are released into atmosphere. Sulphur is cooled in a spray chamber and collected for sale.

H_2S is a very toxic gas produced during refining crude petroleum. Generally, hydrogen sulphide waste gas streams are treated by the Claus process, which is based on partial oxidation of hydrogen sulphide into sulphur and water. The Claus oxidation process requires a suitable oxidant mixture. The Kurchatov Institute in Moscow, Russia, developed a process for the decomposition of hydrogen sulphide into hydrogen and sulphur by applying a microwave plasma (plasmatron). The Argon National Laboratory (ANL) of the USA developed a 'plasma-chemical waste treatment process' in which a hydrogen sulphide waste stream is passed through a microwave-generated plasma reactor where it decomposes into hydrogen and sulphur. ANL test results indicated that this decomposition ranged from 65% to 80% per single pass. Preliminary energy and economic analysis data suggest that the plasma-chemical waste treatment process has the potential for annual energy savings of 40 to 70 trillion Btu or $500 to $1000 million for the refining industries. Further details of the process are available from ANL (Harkness, 1994[79]).

Steel making furnaces generate metallic dust (Electric arc furnaces ,EAF), which use galvanized scrap metals, generate dust which often contains water leachable lead (Pb), cadmium (Cd), chromium (Cr) and zinc (Zn). This kind of dust is classified hazardous and needs to be treated prior to disposal. Currently, combined EAF dust production in Canada and the USA is 677,000 tons per year (Ionescu et al., 1997[80]). A current dust treatment process becomes economical if the treatment scale is 40,000 tons/year or above (Xia and Pickles, 1996[16]).

A large number of EAF mills are mini-mill type operations which need a treatment process that is a small scale, on-site and economic. Ghoreshy and Pickles, 1994[81], Chose microwave energy (900 W, 2450 MHz) for the heating a typical EAF dust mixed with powdered carbon for various length of time. Over 90% zinc was volatilized as ZnO (zinc oxide), which was condensed and collected on an alumina plate placed on top of the reaction crucible. The laboratory scale test results demonstrated that zinc removal was rapid and selective. The iron rich residue can be recycled in an iron or steel making furnace. Steel making slag usually contains 20 wt. % iron. To modify the physical characteristics of and to recover iron from the slag. Hatton and Pickles, 1994[82], Conducted laboratory scale microwave heating tests (1000 W, 2450 MHz). The heating behavior of the steel making slag was investigated with and without the addition of carbon or magnetite. Test results demonstrated that both carbon and magnetite addition increased the heating rate of the slag; 1000°C with carbon, 800°C with magnetite, compared to 650°C without any addition. The amount of iron recovered increased with heating time and reached as high as 90%. Microwave heating altered the physical and chemical properties of the slag.

2.12. Latest developments in microwave processing of minerals ores

The mechanical size reduction of solids is an energy intensive and highly inefficient process. Therefore, there is great incentive to improve the efficiency of size reduction and mineral separation processes. Over several decades, this has promoted significant amounts of research, unfortunately, this has only led to small, incremental improvements in efficiency. One area, which has shown significant promise for improving the efficiency of mineral comminution and separation processes, is microwave assisted grinding.

Until recently, the majority of test work carried out concerning microwave treatment of minerals utilized standard multi-mode cavities, similar to that found in a conventional kitchen microwave oven. The multimode cavity. whilst mechanically simple suffers from poor efficiencies and low electric field strengths, vital to high power adsorption. Whilst the influence of microwave energy from this type of cavity has been shown to have a significant influence on ores and minerals, the inefficiencies of the application method have led to conclusions that at present, microwave treatment of minerals (despite the numerous process benefits) is not viable.

More recent studies have presented studies describing the influence of high electric field strength microwave energy on minerals and ores. It is well known that microwave power density in a material (or volumetric power absorption) is directly proportional to the square of the electric field strength within the material. Therefore, it was shown that if local electric field strength can be magnified energy adsorption or heating rate can be amplified many times without the use of further energy. In turn, this lead to reduced cavity residence times and reductions in the required microwave energy input per ton. Detailed tests at the University of Nottingham have shown that in cavities with high electric field strengths the microwave energy consumption to achieve a desired reduction in strength can be as little as 2% of that required in previous work.

Investigations have been carried out on several economically important ores utilizing a high electric field strength cavity for microwave treatment. A systematic approach was used in order to establish the influence of applied power level and exposure duration on each ore sample. Assessment was made of the influence of particle size on heating rate (this will have an effect on the delivery and presentation methods). During sample treatment, assessment was made of electrical energy consumption, the efficiencies of the system being calculated by standard methods. To support the test programme, results of numerical finite difference simulations are presented which illustrate the importance of microwave and ore variables on post treatment ore properties. The results of the simulations showed that if the microwave energy can be supplied to the sample very rapidly (in order of microseconds) then thermal conduction from the heated phase into the bulk ore can be minimised and thermally induced stress is maximised.

In order to validate the simulation predictions a series of experiments are reported which utilise a pulsed microwave energy delivery system on several ore types. Samples were exposed in a multimode cavity connected to a high voltage solid state modulator and pulse generator. The experimental set up was able to deliver pulses at applied power levels of 1-5MW and pulse durations of 1-4μs at frequency of 2.8GHz.

Scanning electron microscopy and image analysis were used to map the pattern of induced fracture across the pulsed treated samples. It was shown that peak power level was the major influence on the degree of fracture with the highest powers giving greater effects. It was also shown that the fracture induced was predominately grain boundary related and the fractures did not seem to run into each other causing weakening of the bulk ore structure as found in samples treated in continuous wave microwave systems. Treated and untreated samples were then processed by appropriate separation techniques to determine if more valuable mineral could be recovered as a result of treatment. It is shown that pulsed treatment positively influences the recovery of valuable minerals from the different ore types investigated.

3. Summary and recommendations

The information compiled in this chapter demonstrates that microwave energy has the potential for application in mineral treatment and metal recovery processes such as heating, drying, grinding, leaching, roasting, smelting, carbothermic reduction of oxide minerals, pretreatment of refractory gold concentrate or ore, spent carbon regeneration and waste management. Usually, microwave energy is more expensive than electrical energy, mainly due to the low conversion efficiency from electrical energy (50% for 2450 MHz and 85% for 915 MHz). However, the efficiency of microwave heating is often much higher than conventional heating and overcomes the cost of the energy. Generally, mineral processing industries treat a large tonnage of ore or concentrate per day (several thousand to over 30,000 tonnes). Currently, the highest microwave power generator available is 75 kW at 915 MHz. To treat such a large tonnage of ore or concentrate a number of generators would have to be operated in parallel, which may not offer a cost advantage over the conventional process. However, for high value product recovery or low tonnage material treatment microwave energy can offer a cost advantage over the conventional process, for example, pretreatment of refractory gold concentrate, regeneration of CIP spent carbon, roasting or smelting of ore or concentrate for smaller operations. Furthermore, it is possible to apply microwave energy to the leaching of ore or concentrate in slurry or semi-solid mixture (paste) at ambient pressure to yield a metal extraction comparable with pressure leaching. Today's processing industries, including mineral processing, are facing increasing global competition, more stringent environmental regulations, higher overhead costs and shrinking profit margins. The processing industries are addressing these problems in various ways, and their processes are approaching peak product yield as well as performance and productivity efficiency. In the foreseeable future processing industries will be looking for high performance conventional as well as nonconventional processing technology. This is the point at which processes based on microwave energy will get favorable consideration. The continued development of high power microwave generator and precision temperature measuring devices for high temperature operation should have positive impact in the acceptance of microwave assisted mineral treatment process. The current R&D status indicates that microwave energy has the potential to play an important and possibly crucial role in future mineral treatment processes. However, challenges remain to be overcome through a fundamental understanding of microwave interaction with minerals, innovation, R&D investigations and advanced engineering, especially in designing efficient applicators, processes and process control devices.

Author details

S.M. Javad Koleini
Tarbiat Modares University, Iran

Kianoush Barani
Lorestan University, Iran

4. References

[1] Meyer, C., Umm Fawakhir, Bir., 1997, Insights into Ancient Egyptian Mining, JOM, 49 (3), 1997, pp. 64-68.
[2] Oldfather, C.H., 1967, Diodorus of Sicily, Cambridge, Harvard U. Press.
[3] The Tech., Vol. V, Massachusetts Institute of Technology, Boston, April 1, 1886, No. 12, pp. 179-180.
[4] Cowen, R., Exploiting the Earth April, 1999, (online draft copy), http://www.geology.ucdavis.edu/~cowen/~GEL11.
[5] Fitzgibbon, K.E. and Veasey, T.J., 1990, Thermally Assisted Liberation - A Review, Minerals Engineering, vol 3, 1/2, pp 181-185.
[6] Yates, A., Effect of Heating and Quenching Cornish Tin Ores before Crushing, Trans IMM, 28, 1918-1919, pp. 41.
[7] Holman, B.W., 1926, Heat Treatment as an Agent in Rock Breaking, Trans IMM, 26, pp. 219.
[8] Myers, W.M., 1925, Calcining as an Aid to Grinding, J. Am. Ceram. Soc., 8, pp. 839.

[9] Prasher, C.L., 1987, Crushing and Grinding Process Handbook, John Wiley & Sons Limited, Chichester.

[10] Wills, B.A.; Parker, R.H.; Binns, D.G., 1987, Thermally Assisted Liberation of Cassiterite, Minerals and Metallurgical Processing, pp 94-96.

[11] Scheding, W.M., Sherring, A.J., Binns, D., Parker, R.H., Wills, B.A.,1981,The Effect of Thermal Pre-treatment on Grinding Characteristics, Camborne School of Mines Journal, vol 81, pp. 43-44.

[12] Kanellopoulos, A.; Ball, A., The Fracture and Thermal Weakening of Quartzite in Relation to Comminution, Journal of the South African Institute of Mining and Metallurgy, pp 45-52.

[13] Pocock, J., Veasey, T.J., Tavares, L.M., King, R.P.,1975, The Effect of Heating and Quenching on the Grinding Characteristics of Quartzite, Powder Technology, vol. 95, 1998, pp 137-142.

[14] Sherring, A.J., 1981, The Effect of Thermal Pretreatment on Grinding Characteristics, Internal Report, Camborne School of Mines.

[15] Manser, R.J., 1983, The Economics of Thermal Pretreatment, Internal Report, Camborne School of Mines.

[16] Xia, K., Pickles, C.A., 1997, Applications of Microwave Energy in Extractive Metallurgy, a Review, CIM Bulletin, Vol. 90, No. 1011.

[17] Ford, J.D., Pei, D.C.T., High Temperature Chemical Processing via Microwave Absorption, J. Microwave Power, Vol. 2, No. 2, pp. 61-64.

[18] Wong, D., 1975, Microwave Dielectric Constants of Metal Oxides at High Temperature, MSc. Thesis, University of Alberta, Canada.

[19] Tinga, W.R., 1988, Microwave dielectric constants of metal oxides, part 1 and part 2. Electromagnetic Energy Reviews 1 (5), 2–6.

[20] Tinga, W.R., 1989, Microwave dielectric constants of metal oxides, part 1 and part 2. Electromagnetic Energy Reviews 2 (1), 349–351.

[21] Chen, T.T., Dutrizac, J.E., Haque, K.E., Wyslouzil, W., Kashyap, S., 1984, The relative transparency of minerals to microwave radiation. Can. Metall. Quart. 23 (1), 349–351.

[22] McGill, S.L., Walkiewicz, J.W., 1987, Applications of microwave energy in extractive metallurgy. J. Microwave Power and Electromagnetic energy, 22 (3), 175–177.

[23] Walkiewicz, J.W., Kazonich, G., McGill, S.L., 1988, Microwave heating characteristics of selected minerals and compounds. Mineral and Metallurgical Processing, 5 (1), 39–42.

[24] Chunpeng, L., Yousheng, X., Yixin, H., 1990, Application of microwave radiation to extractive metallurgy. Chin. J. Met. Sci. Technol. 6 (2), 121–124.

[25] Barani, K., Koleini. S.M.J., 2012, Effect of sample Geometry and Placement on Iron Ore Processing by Microwave, ICKEM II, Singapore.

[26] Walkiewicz, J.W., Clark, A.E., McGill, S.L., 1991, Microwave-assisted grinding. IEEE Transactions on industry applications 27 (2), 239.

[27] Walkiewicz, J.W., Lindroth, D.P., Clarck, A.E., 1993, Grindability of Taconite Rod Mill Feed Enhanced by Microwave Induced Cracking, SME Annual Meeting, Reno, Nevada.

[28] Tavares, L.M., King, R.P., 1996, Effect of Microwave-Induced Damage on Single-Particle Comminution of Ores, SME Annual Meeting, Phoenix, Arizona.

[29] Tavares, L.M., King, R.P.,1996, Fracture Energies of Copper Ore Subject to Microwave Heating, Utah Comminution Center, University of Utah.

[30] Marland, S., Han, B., Merchant, A., Rowson, N., 2000, The Effect of Microwave Radiation on Coal Grindability, Fuel, 79, pp. 1283-1288.

[31] Kingman, S.W.; Vorster, W.; Rowson, N.A., 2000, The Influence of Mineralogy on Microwave Assisted Grinding, Minerals Engineering, 13 (3), 313-327.

[32] Wang, Y., Forssberg, E., 2000, Microwave Assisted Comminution and Liberation of Minerals, Mineral Processing on the Verge of the 21st Century, Özbayoğlu et al. (eds), Balkema, Rotterdam.

[33] Vorster, W.; Rowson, N.A.; Kingman, S.W., 2001, The Effect of Microwave Radiation Upon the Processing of Neves Corvo Copper Ore, Int. J. Miner. Process., vol. 63, pp 29-44.

[34] Kingman, S.W., Jackson, K., Cumbane, A., Bradshaw, S.M., Rowson, N.A., Greenwood, R., 2004, Recent Developments in Microwave-Assisted Comminution, Int. J. Miner. Process, 74, pp. 71-83.

[35] Amankwah, R.K., Khan, A.U., Pickles, C.A., Yen, W.T., 2005, Improved Grindability and Gold Liberation by Microwave Pretreatment of a Free-milling Gold Ore, Trans. Inst. Min. Metall. C., Vol. 114, pp. C30-C36.

[36] Koleini, S.M.J., Barani, K., 2008, The effect of microwave radiation upon grinding energy of an iron ore, Microwave Technology Conference, Cape town, South Africa.

[37] Barani. K., Koleini., S.M.J., 2010, The effect of microwave treatment upon an iron Ore comminution, international mining congress, Tehran, Iran.

[38] Koleini, S.M.J., Barani, K., Rezaei, B., 2012, The effect of microwave treatment upon dry grinding kinetics of an iron ore, Mineral Processing and Extractive Metallurgy Review Journal, vol. 33 (2), pp.159-169.

[39] Salsman, J.B.; Williamson, R.L.; Tolley, W.K.; Rice, D.A., 1996, Short-Pulse Microwave Treatment of Disseminated Sulfide Ores, Minerals Engineering, vol 9, no.1, pp 43-54.

[40] Whittles, D.N., Kingman, S.W., Reddish, D.J., 2003, Application of Numerical Modelling for Prediction of the Influence of Power Density on Microwave Assisted Grinding, Int. J. Min. Proc., 68, pp. 71-91.

[41] Jones, D.A., Kingman, S.W., Whittles, D.N., Lowndes, I.S., 2005, Understanding Microwave Assisted Breakage, Minerals Engineering, 18, pp. 659-669.

[42] Florek, I., Lovas, M., Murova, I., 1996, The effect of microwave radiation on magnetic properties of grained iron containing minerals, in: Proceedings of the 31st International Microwave Power Symposium, Boston, USA.

[43] Kingman, S.W., Corfield, G.M., Rowson, N.A., 1999, Effect of microwave radiation upon the mineralogy and magnetic processing of amassive Norwegian ilmenite ore, Magn. Electrical, No. 9. pp.131–148.

[44] Kingman, S.W., Rowson, N.A., 2000, The effect of microwave radiation on the magnetic properties of minerals, J. Microwave Power Electromagn. Energy 35 () 144–150.

[45] Cui, Z., Liu, Q., Etsell, T.H., 2002, Magnetic properties of illmenite, hematite and oilsand mineral after roasting, Miner. Eng. Vol. 15 pp.1121–1129.

[46] Sahyoun, C., Kingman, S.W., Rowson, N.A., 2003, The effect of heat treatment on chalcopyrite, Phys. Sep. Sci. Eng. 12 pp.23–30.

[47] Uslu, T., Ataly,U., Arol, A.I., 2003, Effect of microwave heating on magnetic separa- tion of pyrite, Colloid Surf. A: Physicochem. Eng. Aspects 225, (2003), 161– 167.

[48] Znamenackova, I., Lovas, M., 2005, Modification of magnetic properties of siderite ore by microwave energy, Sep. Purif. Technol. 43 (2005), 169– 174.
[49] Waters, K.E., Rowson, N.A., Greenwood, R.W., Williams, A.J., 2007, Characterising the effect of microwave radiation on the magnetic properties of pyrite, Sep. Purif. Technol. 46(2007)9–17.
[50] Barani. K., Koleini., S.M.J., Rezaei. B., 2011, Magnetic properties of an iron ore sample after microwave heating magnetic paper, Sep. Purif. Technol. 76 () 331-336.
[51] Haque, K.E., 1987a. Gold leaching from refractory ores—literature survey. Min. Proc. Extv. Metall. Rev. 2, 235–253.
[52] Haque, K.E., 1987b. Microwave irradiation pretreatment of a refractory gold concentrate. In: Salter, R.S., Wyslouzil, D.M., McDonald, G.W. _Eds.., Proc. Int. Symp. on Gold Metallurgy. Winnipeg, Canada, pp.327–339.
[53] Woodcock, J.T., Sparrow, G.J., Bradhurst, D.H., 1989, Possibilities for using microwave energy in the extraction of gold, Proc. 1st Australian Symp. on microwave power applications. Wollongong, Australia.
[54] Al-Harahsheh, M., Kingman, S., Hankins, N., Somerfield, C., Bradshaw, S., Louw, W., 2005, The Influence of Microwaves on the Leaching Kinetics of Chalcopyrite, Minerals Engineering, 18(13- 14), 1259-1268.
[55] Amankwah, R.K., Khan, A.U., Pickles, C.A., Yen, W.T., 2005, Improved Grindability and Gold Liberation by Microwave Pretreatment of a Free-milling Gold Ore, Trans. Inst. Min. Metall. C., Vol. 114, March, pp. C30-C36.
[56] Nanthakumar, B., Pickles, C.A., Kelebek, S., 2007, Microwave treatment a double refractory gold ore, Minerals Engeneering, Vol.120, pp.1109-1119
[57] Amankwah, R.K., Pickles, C.A., 2008, Microwave roasting of a carbonaceous sulphidic gold, MEI Conference, Microwave Technology, Cape Town, South Africa
[58] Ma,S., Luo, W., Mo, W., Su, X., Liu, P., Yang, J., 2008, Desulfur and de-arsenic of refractory gold concentrate by microwave roasting, MEI Conference, Microwave Technology, Cape Town, South Africa.
[59] Pickles, C.A., 2004, Microwave heating behaviour of nickeliferrous limonitic laterite ores, Minerals Engineering, Vol.17, pp.775-784.
[60] Cumbanea, A.J., Miles, N.J., Lester, E., Kingman, S.W., Bradshaw, S.M., 2008, Dielectric properties of sulphide minerals, MEI Conference, Microwave Technology, Cape Town, South Africa.
[61] Standish, N., Worner, H., 1991, Microwave application in the reduction of metal oxides with carbon. Iron and Steel Maker 18 (5), 59–61.
[62] Gomez, I., Aguilar, J.A., 1995, Microwave for reduction of iron ore pellet by carbon. Mat. Res. Soc. Proc. 366, 347–352.
[63]Standish, N., Worner, H.K., Gupta, G., 1990, Temperature distribution in microwave heated iron ore–carbon composites. J. Microwave Power Electromagnet Energy 25 (2),75–80.
[64] Chunpeng, L., Yousheng, X., Yixin, H., 1990, Application of microwave radiation to extractive metallurgy. Chin. J. Met. Sci. Technol. 6 (2), 121–124.
[65] Cook, N.P., 1986, Microwave Principles and System, Chapter 1. Prentice Hall.
[66] Schiffmann, R.F., 1987, Microwave and dielectric drying. In: Majumder, A.S. (Eds)., Handbook of Industrial Drying. Marcel Dekker, pp. 327–356.

[67] Doelling, M.K., Jones, D.M., Smith, R.A., Nash, R.A., 1992, The development of a microwave fluid-bed processor: 1. Construction and qualification of a prototype laboratory unit, Pharmaceutical Research 9 (11), 1487–1501.

[68] Haque, K.E., 1998, Unpublished data, CANMET, 555 Booth St., Ottawa, ON, K1A 0G1, Canada.

[69] George, C.E., Rao, G.V.N., Thalakola, V., 1994, Thermal desorption of contaminants using microwave heated rotary mixture, Proceedings, 29th Microwave Power Symposium. Chicago, IL.

[70] Matthes, S.A., Farrell, R.F., MacKie, A.J., 1983, A microwave system for the acid dissolution of metal and mineral samples, US Bureau of Mines, TRP 120.

[71] Kingston, H.M., Jassie, L.B., 1985, Introduction to microwave sample preparation, theory and practice, Chapters 2 and 3, ACS professional reference book, Am. Chem. Soc.

[72] Kruesi, P.R., Frahm, V.H., 1982, Process for the recovery of nickel, cobalt and manganese from their oxides and silicates, Canadian patent 1,160,057, US patent 4,311,520 and 4,324,582.

[73] Kruesi, W.H., Kruesi, P.R., 1986, Microwave in laterite processing. Proceedings of CIM 25th Conf. of Metallurgists, Toronto.

[74] Weian, D., 1997, Leaching behaviour of complex sulphide concentrate with ferric chloride by microwave irradiation, Rare Metals 16 (2), 152–155.

[75] Antonucci, V., Correa, C., 1995, Sulphuric acid leaching of chalcopyrite concentrate assisted by application of microwave energy, Proc. of the Copper 95-Cobre 95, Int. Conf. Vol. 111, Santiago, Chile.

[76] Avraamides, J., Miovski, P., Van Hooft, P., 1987, Thermal reactivation of carbon used in the recovery of gold from cyanide pulps and solutions, Research and Development in Extractive Metallurgy, The Aus. I.M.M, Adelaide Branch.

[77] Bradshaw, S.M., Van Wyk, E.J., deSwardt, J.B., 1997, Preliminary economic assessment of microwave regeneration of activated carbon for the carbon in pulp process, J. Microwave Power and Electromagnetic Energy 32 (3), 131–144.

[78] Wicks, G.D., Clark, D.E., Schulz, R.L., Folz, D.C., 1995, Microwave technology for waste management applications including disposition of electronic circuitry, In: Clark, D.E., Folz, D.C., Oda, S.J., Silberglitt, R. (Eds)., Microwaves: theory and application in material processing. Vol. 111, pp. 79–89.

[79] Harkness, J.B.L., 1994, Plasma-chemical waste-treatment process for hydrogen sulphide. Proc. 29th Microwave Power Symposium, Chicago, IL.

[80] Ionescu, D., Meadowcraft, T.R., Barr, P.V., 1997, Classification of EAF dust and ZnO content and an assessment of leach performance, Can. Metall. Quart. 36 (4), 269–281.

[81] Ghoreshy, M., Pickles, C.A., 1994, Microwave processing of electric arc furnace dust, Proc. of 52nd Elect. Furn. Conf., Nashville, TN, pp. 187–195.

[82] Hatton, B.D., Pickles, C.A., 1994, Microwave treatment of ferrous slag. Proc. of Steel making Conf. ISS-AIME, Chicago, USA, Vol. 72, pp. 435–442.

Chapter 2

Applications of Microwave Heating in Agricultural and Forestry Related Industries

Graham Brodie

Additional information is available at the end of the chapter

1. Introduction

Microwave frequencies occupy portions of the electromagnetic spectrum between 300 MHz to 300 GHz. The full range of microwave frequencies is subdivided into various bands (Table 1). Because microwaves are also used in the communication, navigation and defence industries, their use in thermal heating is restricted to a small subset of the available frequency bands. In Australia, the commonly used frequencies include 434 ± 1 MHz, 922 ± 4 MHz, 2450 ± 50 MHz and 5800 ± 75 MHz [1]. These frequencies have been set aside for Industrial, Scientific and Medical (ISM) applications. All these frequencies interact to some degree with moist materials.

Band Designator	Frequency (GHz)	Wavelength in Free Space (centimetres)
L band	1 to 2	30.0 to 15.0
S band	2 to 4	15 to 7.5
C band	4 to 8	7.5 to 3.8
X band	8 to 12	3.8 to 2.5
Ku band	12 to 18	2.5 to 1.7
K band	18 to 27	1.7 to 1.1
Ka band	27 to 40	1.1 to 0.75
V band	40 to 75	0.75 to 0.40
W band	75 to 110	0.40 to 0.27

Table 1. Standard Radar Frequency Letter-Band Nomenclature (IEEE Standard 521-1984)

The major advantages of microwave heating are its short startup, precise control, and volumetric heating [2]. In industry, microwave heating is used for drying [2-5], oil extraction from tar sands, cross-linking of polymers, metal casting [2], medical applications [6], pest control [7], enhancing seed germination [8], and solvent free chemistry [9]. Microwave heating has been applied to various agricultural and forestry problems and products since the 1960's [10]. Studies have been undertaken to use microwave energy: to improve crop handling, storage and preservation; to provide pest and weed control for agricultural production, for food preservation and quarantine purposes; and for preconditioning of products for better quality and more energy efficient processing. This chapter is concerned with microwave heating applications in the agricultural and forestry industries for purposes other than human food processing and consists of a review, update, and discussion of some potential applications that may be of interest to the microwave power and agricultural industries.

2. A brief review of microwave heating in moist materials

Most agricultural and forest products are a heterogeneous mixture of various organic molecules and water, arranged in various geometries. There are some important features of microwave heating that will determine the final temperature and moisture distribution during microwave processing.

Any realistic study of microwave heating in moist materials must account for simultaneous heat and moisture diffusion through the material. The coupling between heat and moisture is well known but not very well understood [11]. Henry [12] first proposed the theory for simultaneous diffusion of heat and moisture in a textile package. Crank [13] later presented a more thorough development of Henry's work. Since then, this theory has been rewritten and used by many authors [11, 14-17]. Microwave heating can be described by a combined heat and moisture diffusion equation that includes a volumetric heating term associated with the dissipation of microwave energy in the material [17]:

$$\nabla^2(pM_v+nT)-\frac{\partial}{\partial t}\left\{\begin{matrix}\left[\frac{1}{\tau_v D_a}\left(1+\frac{(1-a_v)\sigma\rho_s}{a_v}\right)-\frac{n\rho\sigma L}{pk}\right]pM_v+\\ \left[\frac{C\rho}{k}\left(1+\frac{\omega L}{C}\right)-\frac{p(1-a_v)\omega\rho_s}{n\tau_v D_a a_v}\right]nT\end{matrix}\right\}+\frac{nq}{k}=0 \tag{1}$$

This can be expressed in a simpler form if $\Omega = pM_v + nT$:

$$\nabla^2\Omega-\frac{1}{\gamma}\frac{\partial\Omega}{\partial t}+\frac{nq}{k}=0.. \tag{2}$$

The constants of association, p and n, are calculated to satisfy:

$$\frac{1}{\gamma}=\left[\frac{1}{\tau_v D_a}\left(1+\frac{(1-a_v)\sigma\rho_s}{a_v}\right)-\frac{n\rho\sigma L}{pk}\right]=\left[\frac{C\rho}{k}\left(1+\frac{\omega L}{C}\right)-\frac{p(1-a_v)\omega\rho_s}{n\tau_v D_a a_v}\right] \tag{3}$$

Essentially, the combined heat and moisture diffusion coefficient (γ) has two independent values, implying that heating and moisture movement occurs in two independent waves. The slower wave of the coupled heat and moisture system is always slower than either the isothermal diffusion constant for moisture or the constant vapour concentration diffusion constant for heat diffusion, whichever is less, but never by more than one half [12, 13]. The faster wave is always many times faster than either of these independent diffusion constants.

Considerable evidence exists in literature for rapid heating and drying during microwave processing [5, 18]; therefore it is reasonable to assume that the faster diffusion wave dominates microwave heating in moist materials whereas the slower wave dominates conventional heating. A slow heat and moisture diffusion wave should also exist during microwave heating; however observing this slow wave during microwave heating experiments may be difficult and no evidence of its influence on microwave heating has been seen in literature so far.

The fast heat and moisture diffusion wave has a profound effect on biological materials during microwave heating. In particular, very rapid heat and moisture diffusion during microwave heating yields: faster heating compared to conventional heating; and localized steam explosions which may rupture plant and animal cells [18, 19].

Other important phenomena associated with microwave heating include: non-uniform heat and moisture distribution due to the geometry of the microwave applicator [20] and the geometry of the heated material [21]; and phenomenon such as thermal runaway which manifest itself as localised "hot spots" and very rapid rises in temperature [22]. The volumetric heating term (q) in equation (2) is strongly influenced by the geometry of the heated material. Ayappa et al. [2] demonstrate that the equation for electromagnetic power distribution generated in a slab of thickness (W) can be described by:

$$q = \frac{1}{2}\omega\varepsilon_o\kappa''(\tau E)^2\left\{e^{-2\beta z} + \Gamma^2 e^{-2\beta(W-z)} + 2\Gamma e^{-\beta(W-2z)}\cos(\delta + 2\alpha z)\right\} \tag{4}$$

It has been shown elsewhere [21] that using this volumetric heating relationship, the solution for equation (2) is:

$$\Omega(t) = \frac{n\omega\varepsilon_o\kappa''(\tau E)^2}{8k\beta^2}\left\{e^{4\gamma\beta^2 t} - 1\right\}\left\{e^{-2\beta z} + \left(\frac{h}{k} + 2\beta\right) z e^{\frac{-z^2}{4\gamma t}}\right\}\left(1 + \Gamma^2 e^{-2\beta W}\right) \tag{5}$$

From this it can be deduced that the temperature/moisture profiles in thick slabs and rectangular blocks usually result in subsurface heating where the maximum temperature is slightly below the material surface [23].

The microwave's electric field distribution in the radial dimension of a cylinder can be described by [21]:

$$E = \tau E_o \frac{I_o(\beta r)}{I_o(\beta r_o)} \tag{6}$$

The resulting solution to equation (2) can ultimately be derived [21]:

$$\Omega(t) = \frac{n\omega\varepsilon_o\kappa''\tau^2 E_o^2\left(e^{4\beta^2\gamma t} - 1\right)}{4k\beta^2 I_o(2\beta r_o)}\left[\frac{4\alpha\gamma t}{\left[J_o(\alpha r_o) I_o(\beta r_o)\right]^2} e^{\frac{-r^2}{4\gamma t}} + I_o(2\beta r) + \left\{2\beta I_1(2\beta r_o) + \frac{h}{k} I_o(2\beta r_o)\right\}(r_o - r) e^{\frac{-(r_o - r)^2}{4\gamma t}}\right] \tag{7}$$

The temperature/moisture profiles in small-diameter cylinders, such as a plant stem, usually exhibit pronounced core heating [23, 24]. On the other hand, temperature profiles in large cylinders exhibit subsurface heating, with the peak temperature occurring slightly below the surface [23].

Microwave heating in spheres is similar to that in cylinders. The microwave's electric field distribution in the radial dimension of a cylinder can be described by [21]:

$$E = \tau E_o \frac{j_o(fr)}{j_o(fr_o)} \tag{8}$$

The resulting solution to equation (2) can ultimately be derived [21]:

$$\Omega(t) = \frac{n\omega\varepsilon_o\kappa''\tau^2 E_o^2\left(e^{4\beta^2\gamma t}-1\right)}{k\beta\cdot i_o(2\beta r_o)}\left[\frac{\alpha\gamma t}{\left[j_o(\alpha r_o)i_o(\beta r_o)\right]^2}e^{\frac{-r^2}{4\gamma t}}+\frac{i_o(2\beta r)}{4\beta}+\left\{2\beta\cdot i_1(2\beta r_o)+\frac{h}{k}i_o(2\beta r_o)\right\}\frac{(r_o-r)}{4\beta}e^{\frac{-(r_o-r)^2}{4\gamma t}}\right] \tag{9}$$

This analysis can be used, in conjunction with experimental data, to better understand how microwave heating affects agricultural and forestry products.

3. Crop drying

Many studies have investigated the application of microwave energy to speed up crop and wood drying [25, 26]. Higgins and Spooner [27] investigated alfalfa, which was microwave-dried for 7, 8, 9 or 10 min in a microwave oven, compared with field and convective oven-dried alfalfa. They found no differences in crude protein, *in vitro* dry matter digestibility or acid detergent lignin between the various drying methods. Microwave-dried alfalfa generally retained a higher proportion of the cell-wall constituents (neutral detergent fibre) than did field-dried alfalfa. Microwave dried Alfalfa that was treated for 7 minutes had significantly lower acid detergent fibre values than all other drying treatments.

Adu and Otten [28] studied the kinetics of microwave drying of white beans. They found that microwave drying was a falling rate process. When constant power was absorbed, seed temperature increased rapidly to a maximum value during the initial stages of drying and began to decrease gradually during the latter stages of drying. To maintain a constant drying temperature, the microwave power had to be increased progressively as the moisture content of the beans decreased due to drying.

This is linked to reductions in the dielectric properties of the beans as moisture is removed [29], which reduces the interactions between the microwave fields and the beans. The gradual decrease in seed temperature, when the drying rate decreases, is opposite to what is observed during conventional hot air drying. This may be caused by a progressively increasing heat of desorption during the drying process [30], which is a common phenomenon in hygroscopic solids. Thus, the microwave heating characteristics observed for white beans may apply to other hygroscopic solids, such as soils, wood, and fodder chaff.

Microwave drying is fast, as may be expected from the coupling of heat and moisture transport described earlier. The drying curve (Figures 1 and 3) exhibits a short relatively slow drying period, followed by a much faster almost linear relationship between applied microwave energy and moisture loss. This is followed by a more conventional falling rate drying period (Figure 1); however prolonged microwave treatment at high power leads to a phenomenon known as "thermal runaway", which causes charring (Figure 2). Microwave treatment profoundly affects the germination performance of grains, with any reasonable application of microwave power totally inhibiting grain germination (Table 2). The microwave drying curve can be described by:

$$MC = \left(MC_i - MC_f\right) \times e^{-\left(\frac{Em}{b}\right)^2} + MC_f .. \quad (10)$$

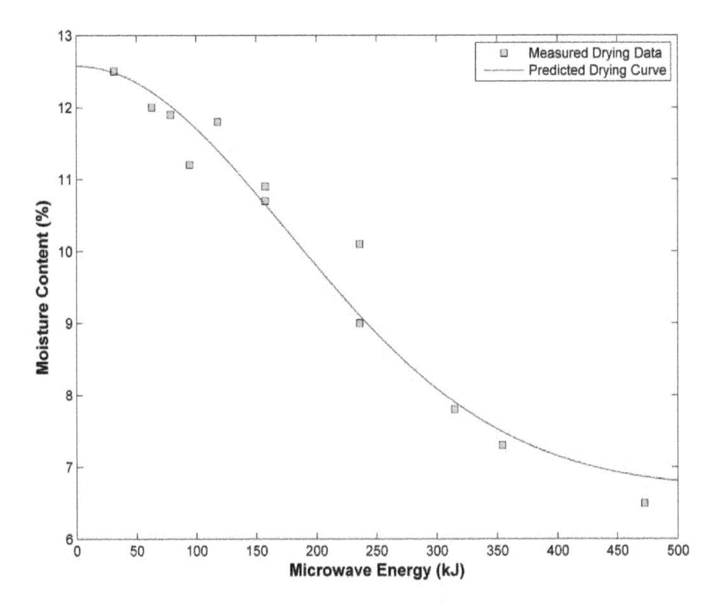

Figure 1. Wheat grain moisture content as a function of applied microwave energy

Figure 2. Charring of wheat during prolonged microwave treatment

Problems with thermal runaway during microwave drying can be overcome by using cyclic drying instead of continuous microwave heating [31]. In this technique, microwave energy is applied for a short time to induce rapid heating and moisture movement and then the product is allowed to equilibrate during a period with no microwave heating. This technique has been successfully applied to timber drying [31] (Figure 3). The resulting drying curve is still described by equation (10).

Microwave Power (%)	Microwave treatment time (minutes)		
	3.5	7.0	10.5
20	88%	42%	4%
50	46%	0%	0%
70	6%	0%	0%
100	0%	0%	0%
Control	90%	88%	94%

Table 2. Effect of microwave drying of 700 g samples of wheat in a 750 W, 2.45 GHz, domestic oven on germination percentages of grains

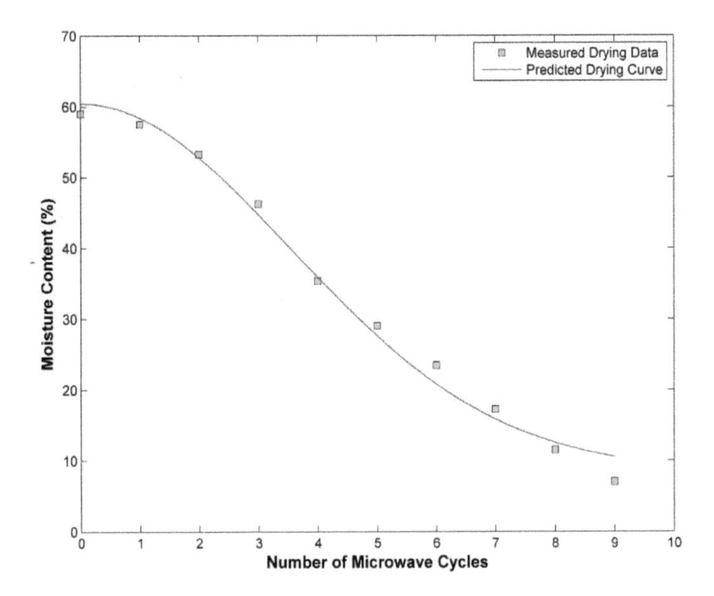

Figure 3. Wood moisture content as a function of the number of microwave energy cycles applied

Walde, et al. [32] studied the effect of microwave drying on the grinding properties of wheat. The microwave dried samples were crisp and consumed less energy for grinding compared to the control samples. The Bond's work index for the bulk sample was 2.26 kWh kg^{-1} compared to 2.41 kWh kg^{-1} for the control samples of equal moisture content. These studies indicated that microwave drying of wheat before grinding helps reduce power consumption in wheat milling. The microwave drying did not change the total protein content, but there were some functional changes in the protein, which was evident from gluten measurements.

Studies have also been carried out on the dry milling characteristics of maize grains, which were dried previously from different initial moisture contents (MCi) in a domestic microwave oven [33]. The MCi ranged from 9.6% to 32.5% on a dry sample basis. Drying was also carried out in a convective dryer at temperatures of 65 – 90 °C. The drying rate curve showed a typical case of moisture loss by diffusion from grains. The dried samples were ground in a hammer mill and the Bond's work index was found to decrease with increasing duration of microwave drying. There was no difference in protein and starch content between the different treatments. Viscosity measurements were made with 10% suspensions of the flour in water which were heated to 80 – 90 °C and allowed to cooled. Viscosity decreased with increasing microwave drying of the grains. The colour analysis showed that flour of the microwave-dried samples was brighter than the control and convective dried samples. Based on these and other studies, microwave drying of agricultural and forestry commodities appear to be a viable alternative to conventional methods, especially when rapid drying and high throughputs of moist material are desirable.

4. Quarantine

Dried timber, nuts and fruits are commonly treated by chemical fumigation to control field and storage pests before being shipped to domestic and international markets. Because chemical fumigants such as methyl bromide are no longer available [34], there is a heightened interest in developing non–chemical pest control. An important key to developing successful thermal treatments is to balance the need for complete insect mortality with minimal impact on the product quality. A common difficulty in using conventional hot–air disinfestation is the slow heating rate, non–uniform temperature distribution, and possible heat damage to heat–sensitive commodities [35]. A more promising approach is to heat the commodity rapidly using radio frequency (RF) or microwave dielectric heating to control insects [35, 36].

Interest in controlling insects, using electromagnetic energy, dates back nearly 70 years. Headlee [37, 38], cites one earlier report of experiments determining lethal exposures for several insect species to 12 MHz electric fields and the body temperatures produced in honey bees due to dielectric heating. Nelson [39] has shown that microwaves can kill insects in grain; however one of the challenges for microwave insect control is to differentially heat the insects in preference to their surrounds. Nelson [39] shows that differential heating depends on microwave frequency. It appears that using a 2.45 GHz microwave system, which is the frequency used in domestic microwave ovens, heats the bulk material, which then transfers heat to the insects; however lower frequencies heat the insects without raising the temperature of the surrounding material beyond 50°C [39].

Nzokou et al [40] investigated the use of kiln and microwave heat treatments for the sanitisation of emerald ash borer (*Agrilus planipennis* Fairmaire) infested logs. Their microwave treatment method was conducted in a 2.8 GHz microwave oven (volume: 0.062 m^3, power: 1250 W) manufactured by Panasonic (Panasonic Co., Secaucus, New Jersey). Due to the limited volume of the microwave oven, two runs were necessary to treat logs assigned to each microwave treatment temperature. Their results showed that a temperature of 65°C was successful at sanitising the infested logs. Microwave treatment was not as effective as kiln treatment, probably because of the uneven distribution of the microwave fields and temperature inside the treated logs. This uneven temperature distribution is partly due to the nature of microwave heating, but may also be due to their choice of microwave chamber used during their experiments.

In spite of this, with the high costs and level of energy needed to thoroughly heat logs to the desired 65 °C using conventional heating, microwave heating is still a very attractive solution for rapid heat sterilisation of infested wood materials [40]. The problem of ensuring appropriate temperature distribution inside treated materials can be easily overcome by using appropriate microwave applicators rather than a multi-mode cavity [41]. Several options are available including conveyer belt feeds through a long choke tunnel into a purpose built applicator [41], or projecting a very intense but short duration microwave field pulse into the material, using an antenna. Plaza *et al.*[42] have developed a system which employs a circular wave-guide energized by two microwave sources oriented at 90 ° to one another. This orthogonal orientation of the microwave fields ensures that they do not interfere with each other, but provides a high power source from relatively cheap mass produced 1 kW magnetrons. Microwave magnetrons of greater power output than 1 kW are usually one or two orders of magnitude more expensive than the 1 kW versions.

It has been shown earlier that microwave heating in moist materials, such as the body of an insect, induces a very fast moving wave of heat and water vapour [17]. The intensity of this wave is directly linked to the intensity of the microwave fields [43], therefore using very intense microwave fields may rupture the internal organs of insects, due to local steam explosions.

The interaction of electromagnetic energy with matter is determined by the dielectric properties of the material. The permittivity of a material can be expressed as a complex quantity, the real part (κ') of which is associated with the capability of the material for storing energy in the electric field of the electromagnetic wave, and the imaginary part (κ'') is associated with the conversion of electromagnetic energy to heat inside the material [39]. This is the phenomenon commonly referred to as dielectric heating. The dielectric properties also determine the reflectivity of a material.

The power dissipated per unit volume in a nonmagnetic, uniform material exposed to radio frequency (RF) or microwave fields can be expressed as:

$$P = (\tau E)^2 \sigma = 55.63 \times 10^{-12} \, f \, (\tau E)^2 \kappa'' \tag{11}$$

Therefore in a system composed of two or more materials, there will be preferential heating in favour of the material with the least reflectivity and higher dielectric loss factor. The rate of temperature increase also depends on the density and thermal capacity of the heated material [35, 36]:

$$\frac{dT}{dt} = \frac{P}{\rho C} \tag{12}$$

Termites are a good example of insects that infest economically important products. For example, in the United States, the annual cost of treating damage caused by the Formosan termite (*Coptotermes formosanus*) exceeds $US 1 billion [44]. The radar cross section of some insect species, including termites, has been modelled by treating them as drops of water of equivalent size and shape [45].

Liquid water exhibits dielectric relaxation at around 22 GHz [46] (Figure 4). There are higher dielectric relaxations in water at about 280 GHz [46], 4.5 THz and 15.4 THz [47]. The dielectric properties of grains, soil and wood also depend on their moisture content [48] (Figure 5).

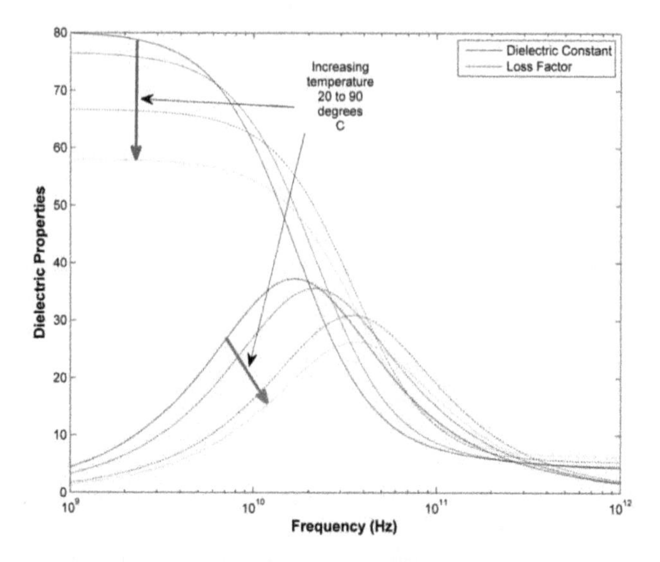

Figure 4. Dielectric properties of pure water as a function of frequency and temperature (calculated using equations and data from literature [46])

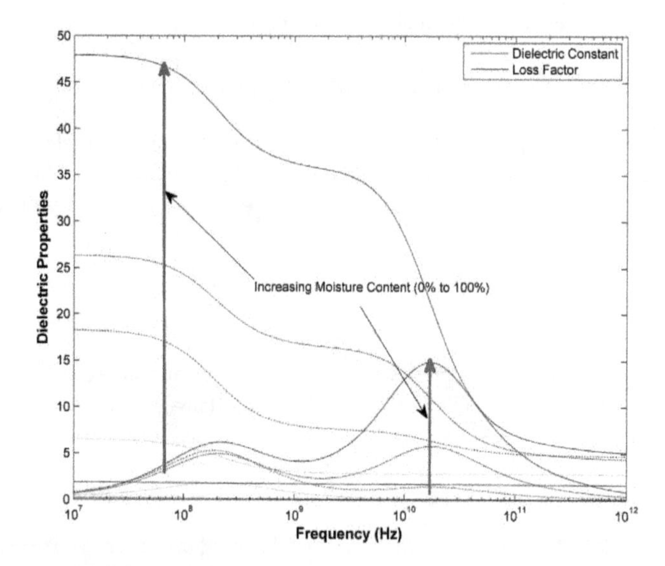

Figure 5. Dielectric properties of wood (density = 500 kg m^{-3}) as a function of frequency and moisture content varying between 0 % and 100 % on a dry wood basis (calculated using equations and data from literature [48])

Dry wood-in-service is in hydro-thermodynamic equilibrium with its surroundings. This condition is known as the equilibrium moisture content [49]. Depending on the atmospheric conditions, equilibrium moisture content is usually about 12% moisture on a dry wood weight for weight basis. When termites invade wood, they often import moisture into the structure to maintain a suitable microclimate for their foraging activities. The maximum moisture content that wood can attain before free water begins to form is known as fibre saturation. This occurs at about 25 - 30 % moisture content [49], depending on the wood species. Fibre saturation refers to the state when all the cells are free of water and only bound water is found within the cell walls. Usually termites do not increase the moisture content beyond fibre saturation. The dielectric properties of termites (modelled as water) and wood at fibre saturation are significantly different from each other (Figure 6).

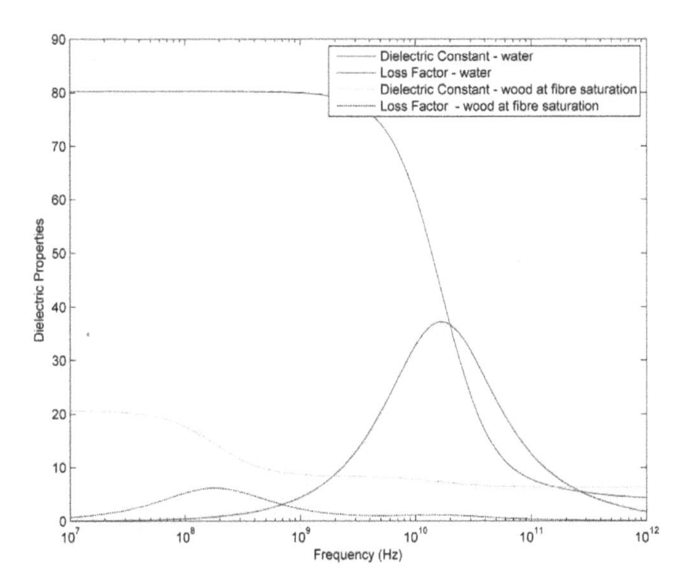

Figure 6. Comparison of the dielectric properties of wood (density = 500 kg m^{-3}) at fibre saturation with water

Treatment of termite infestations using microwave energy, at 2.45 GHz, has been available for some time [50, 51]. This technique does not directly heat the termites, but heats the surrounding wood to more than 55°C [52], which then causes termite mortality. Unfortunately, the combination of high reflectivity and low dielectric losses for water in the lower microwave frequency band (2.45 GHz) means that there is virtually no differential heating between the termites and wood that is at fibre saturation; however significant differential heating should occur once the frequency increases above 20 GHz (Figure 7). Research in the field of ultra-high frequencies (>20 GHz) indicates that these frequencies may selectively heat insect pests in favour of the materials they infest [52, 53]. Therefore research into ultra-high frequency microwave based insect control should yield some valuable insights over the coming decades [52].

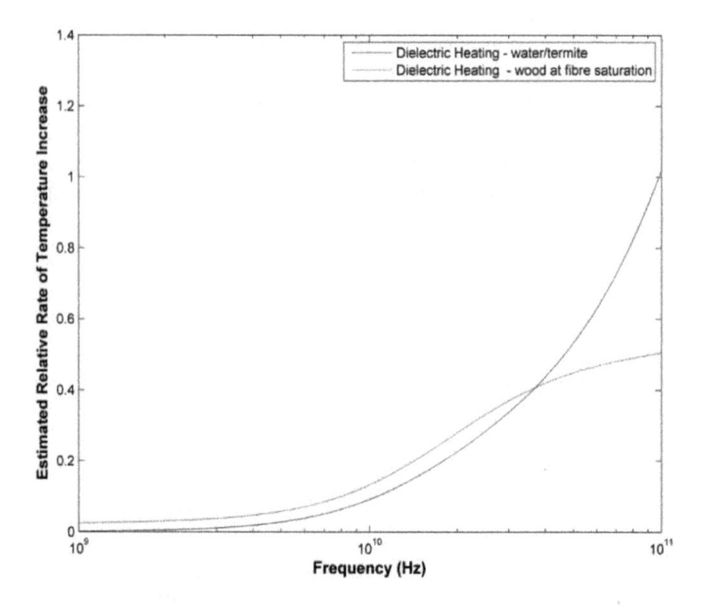

Figure 7. Relative dielectric heating and wood at fibre saturation moisture content, calculated using equation (12) and the dielectric properties of water and wood [48]

Park, *et al.*[54] studied the survival of microorganisms after heating in a conventional microwave oven. Kitchen sponges, scrubbing pads, and syringes were deliberately contaminated with wastewater and subsequently exposed to microwave radiation. The heterotrophic plate count of the wastewater was reduced by more than 99 percent within 1 to 2 minutes of microwave heating. Coliform and E. coli in kitchen sponges were completely inactivated after 30 seconds of microwave heating. Bacterial phage MS2 was totally inactivated within 1 to 2 minutes, but spores of *Bacillus cereus* were more resistant than the other microorganisms tested, requiring 4 minutes of irradiation for complete eradication. Similar inactivation rates were obtained in wastewater-contaminated scrubbing pads; however microorganisms attached to plastic syringes were more resistant to microwave irradiation than those associated with kitchen sponges or scrubbing pads. It took 10 minutes for total inactivation of the heterotrophic plate count and 4 minutes of treatment for total inactivation of total coliform and E. coli. A 4-log reduction of phage MS2 was obtained after 2 minutes of treatment with 97.4 percent reductions after 12 minutes of microwave treatment.

Devine et al. [55] conducted a trial in which microwave radiation, coupled with steam heat, was used to treat organic waste (1,136 kg of culled turkey carcasses), designed to simulate a small-scale poultry mortality event. They inoculated the turkey carcasses with *Bacillus atrophaeus* spores and *Salmonella enterica* before inserting them into a purpose built portable microwave unit (Sanitec Industries), along with other organic waste. The units are designed to treat in excess of 250 kg/per hour of waste. The system has been designed so that the waste is transported through the microwave fields along a screw so that the final exposure time and temperature profile is a minimum of 30 minutes at 95 °C. The system generated a seven-log reduction in the microbial load of Salmonella and a five-log reduction in Bacillus spores. These results illustrate the potential of using microwave radiation for quarantine procedures. The following sections will illustrate more specifically how microwave energy can manage pests in agricultural and forestry systems.

5. Effect of microwave heating on seeds and plants

In 2006, the cost of weed management and loss of production to Australian agricultural industries was estimated to be about $4 billion annually [56]. Depletion of the weed seed bank is critically important to overcoming infestations of various weed species [57]. Mechanical and chemical controls are the most common methods of weed management in cropping systems [58, 59]. The success of these methods usually depends on destroying the highest number of plants during their early growth stages [58] before they interfere with crop production and subsequently set further seed. These strategies must be employed continually to deplete the weed seed bank.

Interest in the effects of high frequency electromagnetic waves on biological materials dates back to the late 19th century, while interest in the effect of high frequency waves on plant material began in the 1920's [60]. In many cases, short exposure of seeds to radio frequency and microwave radiation resulted in increased germination and vigour of the emerging seedlings [61, 62]; however, long exposure usually resulted in seed death [59].

Davis *et al.*[63, 64] were among the first to study the lethal effect of microwave heating on seeds. They treated seeds, with and without any soil, in a microwave oven and showed that seed damage was mostly influenced by a combination of seed moisture content and the energy absorbed per seed. Other findings from their studies suggested that both the specific mass and specific volume of the seeds were strongly related to seed mortality [64]. This could be due to the *"radar cross-section"* [65] presented by seeds to propagating microwaves. Large radar cross-sections allow the seeds to intercept, and therefore absorb, more microwave energy. The geometry of many seeds can be regarded as ellipsoids or even spheres, so the microwave fields are focused into the centre of the seed (see equation 8). Therefore larger seeds focus more energy into their core, which results in higher temperatures at the centre of the seed (see equation 9), leading to higher mortality rates. Seeds whose geometry can be approximated as being cylindrical will also focus more energy into their core as their dimensions increase (see equations 6 and 7).

Barker and Craker [66] investigated the use of microwave heating in soils of varying moisture content (10-280 g water/kg of dry soil) to kill 'Ogle' Oats (*Avena sativa*) seeds and an undefined number of naturalised weed seeds present in their soil samples. Their results demonstrated that a seed's susceptibility to microwave treatment is entirely temperature dependent. When the soil temperature rose to 75°C there was a sharp decline in both oat seed and naturalised weed seed germination. When the soil temperature rose above 80°C, seed germination in all species was totally inhibited.

Several patents dealing with microwave treatment of weeds and their seeds have been registered [67-69]; however none of these systems appear to have been commercially developed. This may be due to concerns about the energy requirements to manage weed seeds in the soil using microwave energy. In a theoretical argument based on the dielectric and density properties of seeds and soils, Nelson [70] demonstrated that using microwaves to selectively heat seeds in the soil *"can not be expected"*. He also concluded that seed susceptibility to damage from microwave treatment is a purely thermal effect, resulting from soil heating and thermal conduction from the soil into the seeds. This has been confirmed experimentally by Brodie *et al.*[71].

Microwaves can kill a range of weed seeds in the soil [63, 64, 72], however fewer studies have considered the efficacy of using microwave energy to manage already emerged weed plants. Davis et al. [63] considered the effect of microwave energy on bean (*Phaseolus vulgaris*) and Honey Mesquite (*Prosopis glandulosa*) seedlings. They discovered that plant aging had little effect on the susceptibility of bean plants to microwave damage, but honey mesquite's resistance to microwave damage increased with aging. They also discovered that bean plants were more susceptible to microwave treatment than honey mesquite plants.

Brodie et al. [73] studied the effect of microwave treatment on Marshmallow (*Malva parviflora*) seedlings, using a prototype microwave system based on a modified microwave oven. The prototype system, energised from the magnetron of the microwave oven operating at 2.45 GHz, has an 86 mm by 43 mm rectangular wave-guide channelling the microwaves from the oven's magnetron to a horn antenna outside of the oven. This allowed the oven's timing circuitry to control the activity of the magnetron.

Horn antennas (Figure 8), like the design used in the experimental prototype are very popular for microwave communication systems [74]. The vertical plane of the horn antenna is usually referred to as the E-plane, because of the orientation of the electrical field (or E-field) in the antenna's aperture. The horizontal plane is referred to as the H-plane, because of the orientation of the magnetic field (or H-field) of the microwave energy. The H-plane electric field distribution in the aperture of a horn antenna, fed from a wave-guide propagating in the TE_{10} mode, is approximated by:

$$E = E_o \cos\left(\frac{\pi}{a}x\right) \tag{13}$$

In the case of a cylindrical object, such as a plant stem, the microwave's electric field distribution created by a horn antenna [21] can be described by:

$$E = \tau E_o \frac{I_o(\beta r)}{I_o(\beta r_o)} \cdot Cos\left(\frac{\pi}{a}x\right) \tag{14}$$

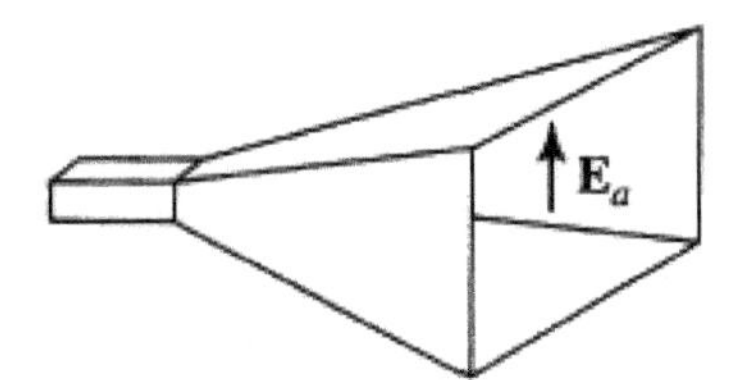

Figure 8. A typical horn antenna showing the orientation of the electrical field component of the microwave energy in the antenna's aperture

The resulting temperature distribution can be described by [75] :

$$T = \frac{n\omega\varepsilon_o\kappa''\tau^2 E_o^2\left(e^{4\beta^2\gamma t}-1\right)}{4k\beta^2 I_o\left(2\beta r_o\right)}\left[\frac{4\alpha\gamma t}{\left[J_o\left(\alpha r_o\right)I_o\left(\beta r_o\right)\right]^2}e^{\frac{-r^2}{4\gamma t}}+I_o\left(2\beta r\right)\right.$$
$$\left.+\left\{2\beta I_1\left(2\beta r_o\right)+\frac{h}{k}I_o\left(2\beta r_o\right)\right\}\left(r_o-r\right)e^{\frac{-\left(r_o-r\right)^2}{4\gamma t}}\right]\cdot Cos\left(\frac{\pi}{a}x\right) \tag{15}$$

Three horn applicators, with varying aperture dimension (180mm by 90 mm; 130 mm by 43 mm; and 86 mm by 20 mm), were developed and tested during various experiments [59, 73, 76, 77]. Aperture size of the horn applicator profoundly affects the treatment time needed to kill plants, with the smaller aperture needing much less time to provide a lethal dose; however the total energy density needed to kill the plants (microwave output power density multiplied by treatment time) was the same irrespective of the horn aperture size. The resulting lethal dose, which was sufficient to kill all the test species, was 350 J cm^{-2}[75] for each plant.

Because energy rather than treatment time is the key factor in plant mortality, two options for using microwave energy to manage weeds become evident; either a prolonged exposure to very diffuse microwave fields or a strategic application of an intensely focused microwave pulse is sufficient to kill plants. Bigu-Del-Blanco, *et al.*[78] exposed 48 hour old seedlings of *Zea mays* (var. Golden Bantam) to 9 GHz radiation for 22 to 24 hours. The power density levels were between 10 and 30 mW cm^{-2} at the point of exposure. Temperature increases of only 4 °C, when compared with control seedlings, were measured in the microwave treated specimens. The authors concluded that the long exposure to microwave radiation, even at very low power densities, was sufficient to dehydrate the seedlings and inhibit their development. On the other hand, recent studies on fleabane (*Conyza bonariensis*) and paddy melon (*Cucumis myriocarpus*)[75, 77] have revealed that a very short (less than 5 second) pulse of microwave energy, focused onto the plant stem, was sufficient to kill these plants. In both cases, rapid dehydration of the plant tissue appears to be the cause of death. This is because microwave heating results in rapid diffusion of moisture [29] through the plant stem as suggested earlier.

Based on energy calculations for plants and seeds on the surface of sandy soil [72, 76], the energy needed to kill dry seeds is an order of magnitude higher than the energy needed to kill already emerged plants. The microwave energy dose needed to kill a paddy melon or fleabane plant was approximately 350 J cm^{-2} (or 35 GJ ha^{-1}) [77]. This is an order of magnitude higher than the embodied energy (2.2 – 3.0 GJ ha^{-1}) associated with chemical weed management [79-82]; however the real microwave energy requirements on a large scale will depend on the plant density and spatial distribution. Therefore the microwave energy requirements may be greatly reduced if appropriate techniques, such as weed seeker systems [83], are employed to only turn on the microwave unit when the system encounters a weed in the field. The growing problems of herbicide resistance [84] also warrants ongoing research and development of microwave weed control technologies. Other strategies may also reduce this energy requirement even further.

It has been well documented that the dielectric properties of most materials are temperature and moisture dependent [22, 85-88]. Ulaby and El-Rayes [89, 90] studied the dielectric properties of plant materials at microwave frequencies. Plants with high moisture content have higher dielectric constants and will therefore interact more with the microwave fields, rendering them more susceptible to microwave damage.

Equation (15) was used in an iterative calculation, where the new dielectric properties for plant based materials were recalculated after every second of microwave heating, based on the changes in temperature and moisture content of the plant during that interval of the microwave heating progresses. This results in non-linear heating responses and sudden jumps in temperature when there is no change in the applied microwave field strength (Figure 9). The sudden jump in temperature for the 15 mm diameter stem is the result of "thermal runaway". The onset of thermal runaway is also dependent on the microwave field intensity (Figure 10) and the heat transfer properties of the heated material (Figure 11). Under the influence of simultaneous heat and moisture diffusion during microwave heating, the effective thermal conductivity of a microwave heated material can be many times the normal value for the plant tissue [17] (see equation (3)).

In most cases, thermal runaway is a problem during microwave heating. It usually leads to undesirable charring of the microwave heated material (see Figure 2) [91]; however it has been very effectively used in some applications such as the development of a microwave drill [92, 93] and preconditioning of wood for further preservative treatment and drying [94, 95].

Vriezinga has concluded that thermal runaway in moist materials, and water in general, is caused by: the specific characteristic of the dielectric properties of water, which decrease with increasing temperature [87] (Figure 4); and resonance of the electromagnetic waves within the irradiated medium due to changes in the dielectric properties of the material during heating [87, 88]. Resonance will only occur when the object's dimensions are some multiple of the wave length of the microwave fields inside the object. That is why thermal runaway only becomes evident in the 15 mm diameter stem (Figure 9), while the smaller stems are too small to allow internal field resonance. Internal steam pressure, induced by

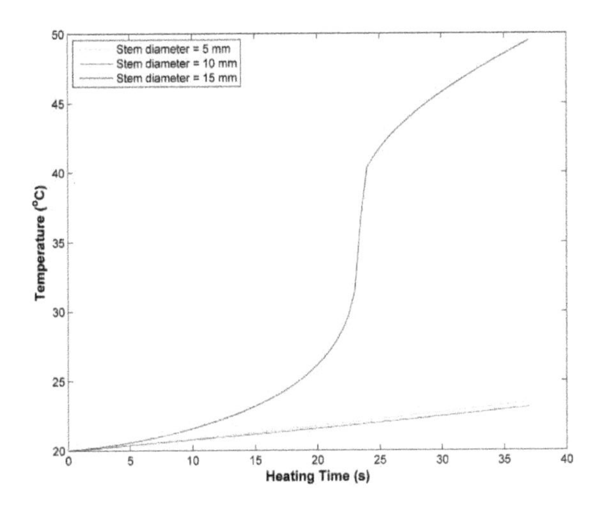

Figure 9. Temperature response, at constant microwave power density at a frequency of 2.45 GHz, in the centre of a plant stem as a function of plant stem diameter, calculated using equation (17) and assuming moisture content loss (MC = 0.87 to 0.10) described by equation (10)

thermal runaway, may cause stem rupture, if sufficient microwave field intensity can be focused onto the plants (Figure 10).

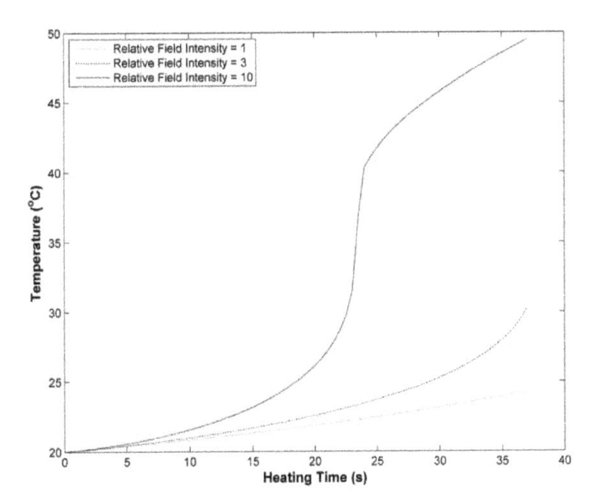

Figure 10. Temperature response in the centre of a 15 mm diameter plant stem as a function of applied microwave field intensity, calculated using equation (17) and assuming moisture content loss (MC = 0.87 to 0.10) described by equation (10)

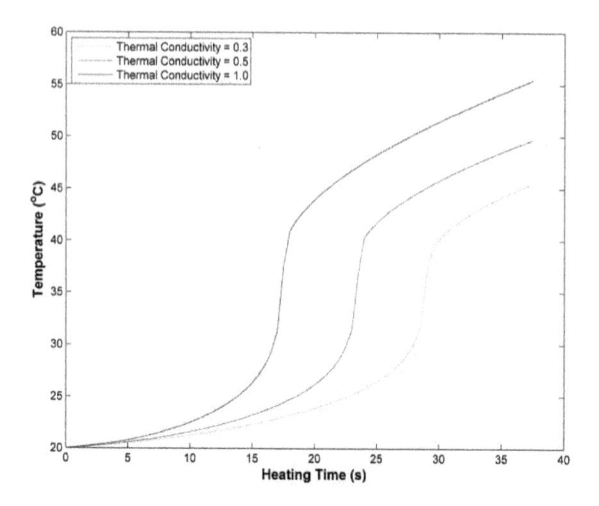

Figure 11. Temperature response in the centre of a 15 mm diameter plant stem as a function of tissue thermal conductivity, calculated using equation (17) and assuming moisture content loss (MC = 0.87 to 0.10) described by equation (10)

Moriwaki et al [96] studied the dehydrochlorination of polyvinyl chloride (PVC) by microwave irradiation using an optical fibre thermo-sensor to investigate the relationship between temperature and microwave absorption onto PVC. Their observations were that: at the beginning of microwave irradiation, the temperature rose in direct proportion to the strength of the incident microwave power and irradiation time; after exceeding a critical condition, the temperature rose quickly (thermal runaway); higher incident microwave power led to thermal runaway starting earlier in the heating process; and higher pre-heating temperatures also led to a faster onset of thermal runaway conditions. These findings are consistent with the modelling displayed in Figures 9, 10, and 11.

Total treatment time, and therefore total applied microwave energy, could be significantly reduced, if thermal runaway can be induced in weed plants during microwave treatment. For example, extrapolating the data presented in Figure 9, it takes approximately 200 to 250 seconds for the 10 mm diameter stem to reach 40 °C; however the 15 mm diameter stem reaches 40 °C in 25 seconds under the same applied microwave power. Therefore the energy required to achieve this temperature rise in the 15 mm diameter stem is only 10 % of the energy needed to heat the 10 mm diameter stem.

Based on existing data, phenomena such as thermal runaway, and the nonlinear temperature/ microwave field strength relationships, it is difficult to discuss "scale up" from small laboratory studies and modelling exercises such as have been used here; however if thermal runaway can be induced in plant tissues, treatment time, and the associated treatment energy, may be drastically reduced; resulting in comparable energy needs to those associated with conventional chemical weed control. This scenario can only be explored by further research into the microwave heating of living plants and plant based materials.

In weed control, microwave radiation is not affected by wind, which extends the application periods compared with conventional herbicide spraying. Energy can also be focused onto individual plants without affecting adjacent plants [75]. This would be very useful for in-crop or spot weed control activities. Microwave energy can also kill the roots and seeds that are buried to a depth of several centimetres in the soil [73, 97].

6. Microwave treatment of animal fodder

Hay is an important feed source for ruminant animals so every effort should be made to improve its feed conversion efficiency and reduce the risk of importing weed seeds as hay is transported from one location to another. Similarly, cereal grains are the base of most horse rations, because they are a valuable source of digestible energy; however their use is always associated with some risk.

The major concern when feeding cereal grains to horses is the risk of incomplete starch digestion in the small intestine, which enables significant amounts of starch to pass through to the caecum and colon. When starch is able to reach these organs it rapidly ferments producing an accumulation of acidic products, which place the horse at risk of developing serious and potentially fatal illnesses such as laminitis, colic and ulcers [98].

Dong *et al.*[99] discovered that organic matter degradability of wheat straw in the rumen of yaks was increased by around 20% after 4 min of treatment in a 750 W, 2.54 GHz, microwave oven. Sadeghi and Shawrang [100] showed that microwave treatment of canola meal increased *in vitro* dry matter disappearance, including substances that were deemed to be ruminally undegradable. Sadeghi and Shawrang [101] also showed that microwave treatment reduced the rumen degradable starch fraction of corn grain and decreased crude protein degradation of soya-bean meal [102] compared with untreated samples. No studies of microwave treatment of horse feeds could be found in the available literature.

Small scale *in vitro* pepsin-cellulase digestion experiments [103], similar to the technique developed by McLeod and Minson [104, 105], demonstrated that microwave treatment: increased dry matter percentage with increasing microwave treatment time; increased *in vitro* dry matter disappearance with increasing microwave treatment time; but had no significant effect on post-digestion crude protein content.

When 25 kg bags of lucerne fodder, treated in an experimental 6 kW, 2.45 GHz, microwave heating chamber [31] were subjected to a similar *in vitro* pepsin-cellulase digestion study, dry matter disappearance significantly increased compared to the untreated samples; however there was no significant difference attributable to the duration of microwave treatment. Feeding 12-14 month old Merino sheep on a "maintenance ration" of microwave treated Lucerne resulted in a significant increase in body weight instead of the relatively constant body weight that would be expected from a maintenance ration. By the end of the 5 week feeding trial the control group was only 0.4 % heavier than when they started, which would be expected from a maintenance ration. However the group being fed the microwave treated lucerne gained 7 % of their initial body weight in the second week of the trial and maintained this body weight until the end of the trial. Their finishing weight after 5 weeks was 8.1 % higher than their starting weight [103].

In vitro assessment of microwave treated oats, using the Megazyme Total Starch Assay Procedure [106], which simulates the initial digestive processes in the stomach and small intestines of a horse, demonstrated significantly increased starch digestion. This implies that less undigested starch should proceed through the intestinal tract where it could cause significant health risks to the animal.

The efficiency of chaff and fodder treatment using microwave energy depends on the applied microwave energy and the frequency at which the microwave system operates. Absorbed energy, calculated by measuring the combination of sensible (temperature rise) and latent heat (moisture loss) in treated samples, is much higher at 2.45 GHz than at 922 MHz (Figure 12). It is also evident that efficiency (i.e. the ratio of absorbed energy to applied microwave energy) decreases as the applied microwave energy increases (Figure 13). This is attributable to the increasing transparency of the fodder material to microwave energy as it dries during microwave treatment. The dielectric properties, and therefore the microwave heating effect, reduce as the moisture content of plant materials decrease (Figure 5). Some of these problems of material transparency during microwave treatment can be overcome by compressing the fodder, which increases its ability to absorb microwave energy (Figure 14).

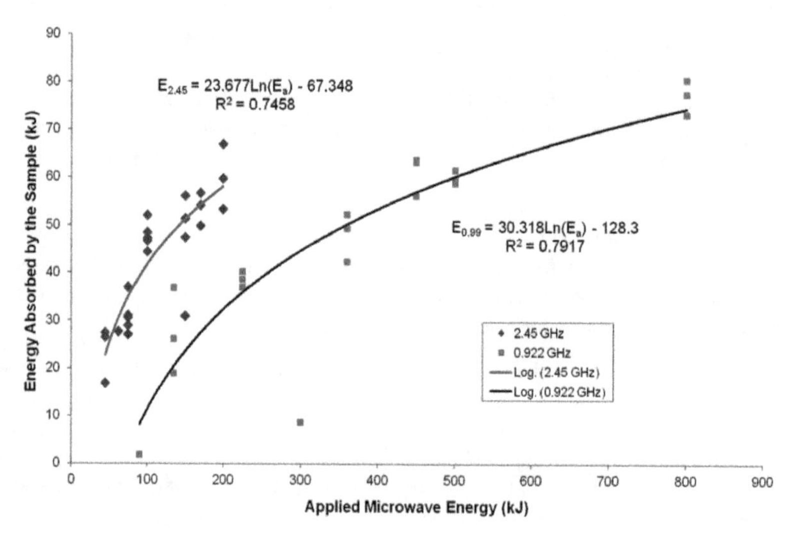

Figure 12. Absorbed energy in crop chaff (fodder) as a function of applied microwave energy for 922 MHz and 2.45 GHz

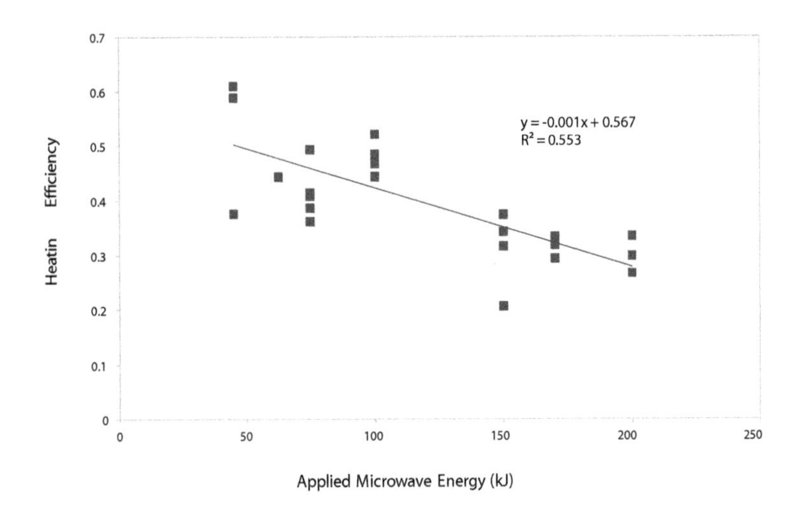

Figure 13. Microwave heating efficiency in crop chaff (fodder) as a function of applied microwave energy at 2.45 GHz

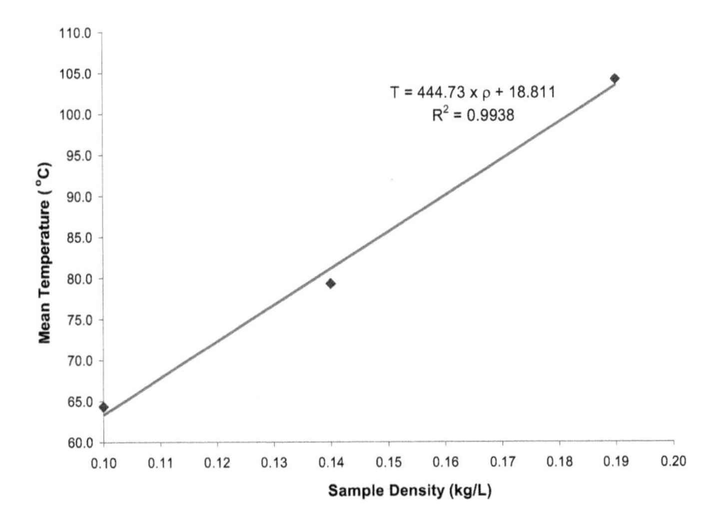

Figure 14. Mean temperature of 500 g samples of microwave treated fodder chaff as a function of material density (ρ) when heated by a 2.5 kW microwave source operating at 2.45 GHz for 30 seconds

Brodie et al. [103] treated 25 kg samples of lucerne chaff in an experimental 6 kW microwave chamber [31]. The temperature in the air space at the top of the lucerne bags rose to 100 °C in ~12 min and fluctuated above 100 °C for the remainder of the treatment time (Figure 15). The maximum temperature in the air space was 115 °C. The maximum temperature in the lucerne (99.5 °C) was measured by the probe facing the microwave magnetrons whereas the maximum temperature measured by the probe in the front of the bag, facing the door of the microwave chamber, was only 94 °C.

The temperature in the lucerne increased steadily at a rate of ~2 °C/min of microwave heating time for the first 20–25 min of heating. At this stage there was a sudden increase in heating rate (~6 °C/min) until the temperature stabilised at ~98 °C (Figure 15). The sudden jump in the heating rate after some time of steady heating may be evidence of thermal runaway (Figure 9). Observation of the treated chaff showed no signs of charring; however the chaff was dry and crisp [103]. The onset of thermal runaway dramatically increases the heating efficiency. In this example, the heating rate during thermal runaway is three times higher than during the normal heating phase. Provided charring can be avoided, inducing thermal runaway in the treated chaff may drastically improve treatment efficiency. The onset of thermal runaway is usually quicker when the microwave field intensity is higher (Figure 10) and the thermal conductivity of the material is increased. In the case of fodder chaff, thermal conductivity is proportional to density, which may partially explain why increasing the material density significantly increases the heating rate of the chaff (Figure 14). This needs further exploration.

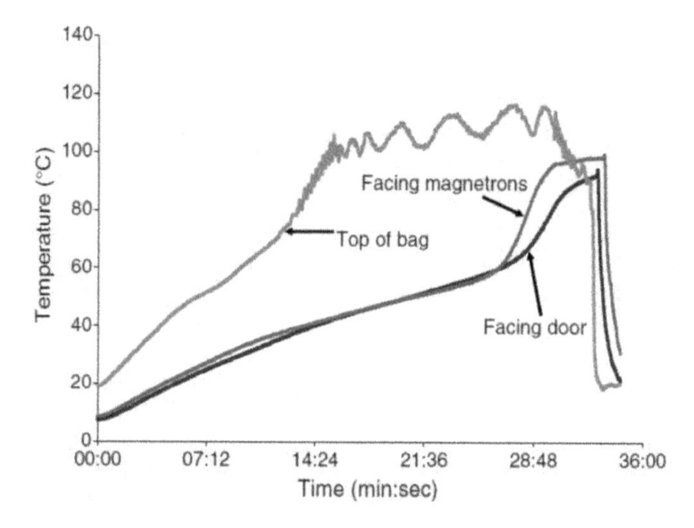

Figure 15. Temperature data from three locations within one 25-kg bag of lucerne being treated for 30 minutes

7. Microwave assisted extraction

During microwave assisted extraction (MAE), plant materials such as wood, seeds and leaves are suspended in solvents and the mixture is exposed to microwave heating instead of conventional heating. Enhanced rates of plant oil extraction have been observed for a range of plant materials. Chen and Spiro [107] examined the extraction of the essential oils of peppermint and rosemary from hexane and ethanol mixtures and found that yields were more than one third greater in the microwave assisted extractions. Saoud *et al.*[108] studied MAE of essential oils from tea leaves and achieved higher yields (26.8 mg/g) than conventional steam distillation (24 mg/g).

Chemat *et al.* [109] studied the extraction of oils from limonene and caraway seeds and found that MAE led to more rapid extraction as well as increased yields. Scanning electron microscopy of the microwave treated and untreated seeds revealed significantly increased rupture of the cell walls in the treated seeds. MAE also led to a more chemically complex extract, which was thought to be a better representation of the true composition of the available oils in caraway seed.

Although less well described in the literature, an alternative approach for utilizing microwave heating of plant based materials has been to treat the materials with microwave energy prior to conventional extraction processes [110]. Microwave preconditioning of sugar cane prior to juice diffusion studies led to significant decreases in colour and significant increases in juice yield, Brix %, purity and Pol % [111]. Microwave treatment significantly reduced the compression strength of the sugar cane samples [111], especially while the cane was still hot from the microwave treatment. This treatment option reduced the compressive strength of the cane to about 18 % of its original strength, implying that much less energy would be required to crush the cane for juice extraction.

Controlled application of microwave heating to green timber [18, 41, 94, 112-114] results in local steam explosions and can directly manipulate both permeability and density with potentially less strength loss than is caused by conventional steam conditioning. This technique does not attempt to dry the wood using microwave energy. Rather, it is used to modify the wood structure to facilitate faster drying in more conventional systems. This technology has the potential to: relieve internal log stresses in susceptible species; substantially accelerate drying; improve preservative treatment and resin uptake; and produce new wood-based products for commercial applications. Application of microwave processing technology has the potential to streamline production and to facilitate conveyor belt automation in the timber industry.

8. Microwave assisted pyrolysis and bio-fuel extraction

Three different thermo-chemical conversion processes are possible, depending on the availability of oxygen during the process: combustion (complete oxidation), gasification (partial oxidation) and pyrolysis (thermo-chemical degradation without oxygen). Among these, combustion is the most common option for recovering energy. Combustion is also associated with the generation of carbon oxides, sulphur, nitrogen, chlorine products (dioxins and furans), volatile organic compounds, polycyclic aromatic hydrocarbons, and dust [115]; however gasification and pyrolysis offer greater efficiencies in energy production, recovery of other compounds and less pollution.

Most studies of pyrolysis behaviour have considered lingo-cellulosic materials, which comprise of a mixture of hemicellulose, cellulose, lignin and minor amounts of other organic compounds. While cellulose and hemicelluloses form mainly volatile products during pyrolysis due to the thermal cleavage of the sugar units, lignin mainly forms char since it is not readily cleaved into lower molecular weight fragments [115]. Wood, crops, agricultural and forestry residues, and sewage sludges [116] can be subjected to pyrolysis processes to recover valuable chemicals and energy.

Conventional heating transfers heat from the surface towards the centre of the material by convection, conduction and radiation; however microwave heating is a direct conversion of electromagnetic energy into thermal energy within the volume of the material [20]. In microwave heating, the material is at higher temperature than its surroundings, unlike conventional heating where it is necessary for the surrounding atmosphere to reach the desired operating temperature before heating the material [115]. Consequently, microwave heating favours pyrolysis reactions involving the solid material, while conventional heating improves the reactions that take place in surroundings, such as homogeneous reactions in the gas-phase [115]. In microwave heating, the lower temperatures in the microwave cavity can also be useful for condensing the final pyrolysis vapours on the cavity walls.

Microwave assisted pyrolysis yields more gas and less carbonaceous (char) residue, which demonstrate the efficiency of microwave energy [115]. The conversion rates in microwave assisted pyrolysis are always higher than those observed in conventional heating at any temperature. The differences between microwave heating and conventional heating seems to be reduced with temperature increase, which points to the higher efficiency of microwave heating at lower temperatures [115].

Bio-fuel extraction is facilitated when microwave energy is used to thermally degrade various organic polymers to facilitate extraction of sugars for fermentation [117]. These sugars can then be fermented and distilled to create fuel alcohols. Woody plant materials are commonly subjected to microwave assisted bio-fuel extraction; however other materials such as discharge from food processing industries, agriculture and fisheries can also be processed using these techniques. Other materials that have been subjected to microwave assisted bio-fuel extraction include: soybean residue; barley malt feed; tea residues; stones from Japanese apricots; corn pericarp, which is a by-product from corn starch production; and Makombu (*Laminaria japonica*), which is a kind of brown sea algae.

9. Conclusion

Microwave and radio frequency heating have many potential applications in the agricultural and forestry industries. This chapter has discussed a few of these, but there are many more that have not been included. The purpose of this chapter was to encourage practitioners within the microwave engineering and agricultural and forestry industries to explore the many possibilities of applying microwave heating to address many problems and opportunities within the primary industries.

Nomenclature

Ω= combined temperature and moisture vapour parameter
a_v= air space fraction in the material
b= Microwave drying constant to be determined experimentally
C= thermal capacity of the composite material (J kg^{-1} $°C^{-1}$)
D_a= vapor diffusion coefficient of water vapor in air (m^2 s^{-1})
E= electric field associated with the microwave (V m^{-1})
E_o= magnitude of the electric field external to the work load (V m^{-1})
Em= Microwave energy (J)
f= complex wave number of the form $f = \alpha + j\beta$
h= convective heat transfer at the surface of a heated object (W m^{-1} K^{-1})
$i_o(x)$ = modified spherical Bessel function of the first kind of order zero
$I_o(x)$= modified Bessel function of the first kind of order zero
$j_o(x)$ = spherical Bessel function of the first kind of order zero
$J_o(x)$ = Bessel function of the first kind of order zero
$i_1(x)$ = modified spherical Bessel function of the first kind of order one
$I_1(x)$= modified Bessel function of the first kind of order one
k= thermal conductivity of the composite material (W m^{-1} $°C^{-1}$)
L= latent heat of vaporization of water (J kg^{-1})
M_v= moisture vapor concentration in the pores of the material (kg m^{-3})
MC= Moisture content (kg kg^{-1} dry matter)
MC_i= Initial moisture content (kg kg^{-1} dry matter)
MC_f= Final moisture content (kg kg^{-1} dry matter)
n = constant of association relating water vapor concentration to internal temperature of a solid
p = constant of association relating internal temperature of a solid to water vapor concentration
q= volumetric heat generated by microwave fields (W m^{-3})
r= radial distance form the centre of a cylinder or sphere (m)
r_o= external radius of the cylinder or sphere (m)
t= heating time (s)
T= temperature (°C)
W= thickness of the slab (m)
x= linear distance across the aperture of a horn antenna (m)
z= linear distance from the surface of a slab (m)
Γ= internal reflection coefficient
α = real part of the complex wave number f
β= imaginary part of the complex wave number f
δ= phase shift of microwave fields at the surface of a material
ε_o= electrical permittivity of free space
γ= combined diffusivity for simultaneous heat and moisture transfer
κ' = relative dielectric constant of the material

κ'' = dielectric loss factor of the material
λ= wave length inside a material (m)
ρ= composite material density (kg m^{-3})
ρ_s= density of the solid material (kg m^{-3})
σ= constant of association relating moisture vapor concentration to moisture content in a solid
τ= transmission coefficient for incoming microwave
τ_v= tortuosity factor
ω= angular frequency (rad s^{-1})

Author details

Graham Brodie
Melbourne School of Land and Environment, University of Melbourne, Australia

10. References

[1] Commonwealth Department of Transport and Communications (1991) Australian Radio Frequency Spectrum Allocations. Canberra: Commonwealth Department of Transport and Communications
[2] Ayappa K G, Davis H T, Crapiste G, Davis E J and Gordon J (1991) Microwave heating: An evaluation of power formulations. Chemical Engineering Science. 46: 1005-1016.
[3] Antti A L and Perre P (1999) A microwave applicator for on line wood drying: Temperature and moisture distribution in wood. Wood Science and Technology. 33: 123-138.
[4] Hasna A, Taube A and Siores E (2000) Moisture Monitoring of Corrugated Board During Microwave Processing. Journal of Electromagnetic Waves and Applications. 14: 1563.
[5] Zielonka P and Dolowy K (1998) Microwave Drying of Spruce: Moisture Content, Temperature and Heat Energy Distribution. Forest Products Journal. 48: 77-80.
[6] Bond E J, Li X, Hagness S C and Van Veen B D (2003) Microwave imaging via space-time beamforming for early detection of breast cancer. IEEE Transaction on Antennas and Propagation. 51: 1690-1705.
[7] Nelson S O (1972) Insect-control possibilities of electromagnetic energy. Cereal Science Today. 17: 377-387.
[8] Nelson S O and Stetson L E (1985) Germination responses of selected plant species to RF electrical seed treatment. Transactions of the ASAE. 28: 2051-2058.
[9] Arrieta A, Otaegui D, Zubia A, Cossio F P, Diaz-Ortiz A, delaHoz A, Herrero M A, Prieto P, Foces-Foces C, Pizarro J L and Arriortua M I (2007) Solvent-Free Thermal and Microwave-Assisted [3 + 2] Cycloadditions between Stabilized Azomethine Ylides and Nitrostyrenes. An Experimental and Theoretical Study. Journal of Organic Chemistry. 72: 4313-4322.

[10] Nelson S O (2003) Microwave and radio-frequency power applications in agriculture. In: Folz D C, Booske J H, Clark D E and Gerling J F editors. Third World Congress on Microwave and Radio Frequency Applications. Westerville, OH: The American Ceramic Society. pp. 331-340

[11] Chu J L and Lee S (1993) Hygrothermal stresses in a solid: Constant surface stress. Journal of Applied Physics. 74: 171-188.

[12] Henry P S H (1948) The diffusion of moisture and heat through textiles. Discussions of the Faraday Society. 3: 243-257.

[13] Crank J (1979) The Mathematics of Diffusion. Bristol: J. W. Arrowsmith Ltd.

[14] Vos M, Ashton G, Van Bogart J and Ensminger R (1994) Heat and Moisture Diffusion in Magnetic Tape Packs. IEEE Transactions on Magnetics. 30: 237-242.

[15] Fan J, Luo Z and Li Y (2000) Heat and moisture transfer with sorption and condensation in porous clothing assemblies and numerical simulation. International Journal of Heat and Mass Transfer. 43: 2989-3000.

[16] Fan J, Cheng, X., Wen, X., and Sun, W. (2004) An improved model of heat and moisture transfer with phase change and mobile condensates in fibrous insulation and comparison with experimental results. International Journal of Heat and Mass Transfer. 47: 2343–2352.

[17] Brodie G (2007) Simultaneous heat and moisture diffusion during microwave heating of moist wood. Applied Engineering in Agriculture. 23: 179-187.

[18] Torgovnikov G and Vinden P (2003) Innovative microwave technology for the timber industry. In: Folz D C, Booske J H, Clark D E and Gerling J F editors. Microwave and Radio Frequency Applications: Proceedings of the Third World Congress on Microwave and Radio Frequency Applications. Westerville, Ohio: The American Ceramic Society. pp. 349-356

[19] Brodie G (2010) Wood: Microwave Modification of Properties. In: Heldman D R editor. Encyclopedia of Agricultural, Food and Biological Engineering. London: Taylor & Francis. pp. 1878 - 1881

[20] Metaxas A C and Meredith R J (1983) Industrial Microwave Heating. London: Peter Peregrinus.

[21] Brodie G (2008) The influence of load geometry on temperature distribution during microwave heating. Transactions of the American Society of Agricultural and Biological Engineers. 51: 1401-1413.

[22] Vriezinga C A (1998) Thermal runaway in microwave heated isothermal slabs, cylinders, and spheres. Journal of Applied Physics. 83: 438 -442.

[23] Van Remmen H H J, Ponne C T, Nijhuis H H, Bartels P V and Herkhof P J A M (1996) Microwave Heating Distribution in Slabs, Spheres and Cylinders with Relation to Food Processing. Journal of Food Science. 61: 1105-1113.

[24] Ohlsson T and Risman P O (1978) Temperature distributions of microwave heating - spheres and cylinders. Journal of Microwave Power and Electromagnetic Energy. 13: 303–310.

[25] Manickavasagan A, Jayas D S and White N D G (2006) Non-Uniformity of Surface Temperatures of Grain after Microwave Treatment in an Industrial Microwave Dryer. Drying Technology. 24: 1559–1567.

[26] Setiady D, Tang J, Younce F, Swanson B A, Rasco B A and Clary C D (2009) Porosity, Color, Texture, and Microscopic Structure of Russet Potatoes Dried Using Microwave Vacuum, Heated Air, and Freeze Drying Applied Engineering in Agriculture. 25: 719-724.

[27] Higgins T R and Spooner A E (1986) Microwave drying of alfalfa compared to field-and oven-drying: Effects on forage quality. Animal Feed Science and Technology. 16: 1-6.

[28] Adu B and Otten L (1996) Microwave heating and mass transfer characteristics of white beans. Journal of Agricultural Engineering Research. 64: 71 -78.

[29] Nelson S O (1987) Frequency, moisture, and density dependence of the dielectric properties of small grains and soybeans. Transactions of the American Society of Agricultural and Biological Engineers. 30: 1538-1541.

[30] Kiranoudis C T, Maroulis Z B, Tsami E and Marinos-Kouris D (1993) Equilibrium moisture content and heat of desorption of some vegetables. Journal of Food Engineering. 20: 55-74.

[31] Harris G A, Brodie G I, Ozarska B and Taube A (2011) Design of a Microwave Chamber for the Purpose of Drying of Wood Components for Furniture. Transactions of the American Society of Agricultural and Biological Engineers. 54: 363-368.

[32] Walde S G, Balaswamy K, Velu V and Rao D G (2002) Microwave drying and grinding characteristics of wheat (Triticum aestivum). Journal of Food Engineering. 55: 271-276.

[33] Velu V, Nagender A, Prabhakara Rao P G and Rao D G (2006) Dry milling characteristics of microwave dried maize grains (Zea mays L.). Journal of Food Engineering. 74: 30-36.

[34] Carter C A, Chalfant J A, Goodhue R E, Han F M and DeSantis M (2005) The Methyl Bromide Ban: Economic Impacts on the California Strawberry Industry. Review of Agricultural Economics. 27: 181-197.

[35] Wang S, Tang J, Cavalieri R P and Davis D C (2003) Differential heating of insects in dried nuts and fruits associated with radio frequency and microwave treatment. Transactions of the ASAE. 46: 1175–1182.

[36] Nelson S O (2001) Radio-frequency and microwave dielectric properties of insects. Journal of Microwave Power and Electromagnetic Energy. 36: 47-56.

[37] Headlee T J (1931) The difference between the effect of radio waves on insects and on plants. Journal of Economic Entomology. 24: 427-437.

[38] Headlee T J and Burdette R C (1929) Some Facts Relative to the Effect of High Frequency Radio Waves on Insect Activity. Journal of the New York Entomological Society. 37: 59-64.

[39] Nelson S O (1996) Review and assessment of radio-frequency and microwave energy for stored-grain insect control. Transactions of the ASAE. 39: 1475-1484.

[40] Nzokou P, Tourtellot S and Kamdem D P (2008) Kiln and microwave heat treatment of logs infested by the emerald ash borer (Agrilus planipennis Fairmaire) (Coleoptera: Buprestidae). Forest Products Journal. 58: 68.

[41] Torgovnikov G and Vinden P (2009) High-intensity microwave wood modification for increasing permeability. Forest Products Journal. 59: 84-92.

[42] Plaza P J, Zona A T, Sanshis R, Balbastre J V, Martinez A, Munoz E M, Gordillo J and Reyes E d l (2007) Microwave disinfestation of bulk wood. Journal of Microwave Power and Electromagnetic Energy. 41: 21-36.

[43] Brodie G I (2008) Innovative wood drying: Applying microwave and solar technologies to wood drying. Saarbruecken, Germany: VDM Verlag. 120 p.

[44] Pimentel D, Lach L, Zuniga R and Morrison D (2002) Environmental and economic costs associated with non-indigenous species in the United States. In: Pimentel D editor. Biological invasions: economic and environmental costs of alien plant, animal, and microbe species. Boca Raton; USA: CRC Press Inc. pp. 285-303

[45] Riley J R (1992) A millimetric radar to study the flight of small insects. Electronics & Communication Engineering Journal. 4: 43-48.

[46] Meissner T and Wentz F J (2004) The Complex Dielectric Constant of Pure and Sea Water From Microwave Satellite Observations. IEEE Transactions on Geoscience and Remote Sensing. 42: 1836-1849.

[47] Ellison W J, Lamkaouchi K and Moreau J M (1996) Water: a dielectric reference. Journal of Molecular Liquids. 68: 171-279.

[48] Torgovnikov G I (1993) Dielectric Properties of Wood and Wood-Based Materials. Berlin: Springer-Verlag.

[49] Jackson A and Day D (1989) Wood Worker's Manual. Sydney: William Collins Sons & Co. Ltd.

[50] Lewis V R and Haverty M I (1996) Evaluation of six techniques for control of the western drywood termite (Isoptera: Kalotermitidae) in structures. Journal of Economic Entomology. 89: 922-934.

[51] Lewis V R, Power, A. B., and Haverty, M. I. (2000) Laboratory evaluation of microwaves for control of the western drywood termite. Forest Products Journal. 50: 79-87.

[52] Tirkel A Z, Lai J C S, Evans T A and Rankin G A (2011) Heating and Provocation of Termites Using Millimeter Waves. Progress in Electromagnetic Research Symposium (Online). 7: 27-30.

[53] Halverson S L, Burkholder W E, Bigelow T S, Nordheim E V and Misenheimer M E (1996) High-power microwave radiation as an alternative insect control method for store products. Journal of Economic Entomology. 89: 1638-1648.

[54] Park D, Bitton G and Melker R (2006) Microbial inactivation by microwave radiation in the home environment. Journal of Environmental Health. 69: 17-24.

[55] Devine A A, Grunden A M, Krisiunas E, Davis D K, Rosario G, Scott S, Faision S and Cosby W M (2007) Testing the Efficacy of a Combination of Microwave and Steam Heat for Log Reduction of the Microbial Load Following a Simulated Poultry Mass Mortality Event. Applied Biosafety. 12: 79-84.

[56] DAFF (2006) Weeds. Australian Department of Agriculture, Fisheries and Forestry

[57] Kremer R J (1993) Management of Weed Seed Banks with Microorganisms. Ecological Applications,. 3: 42-52.

[58] Batlla D and Benech-Arnold R L (2007) Predicting changes in dormancy level in weed seed soil banks: Implications for weed management. Crop Protection/Weed Science in Time of Transition. 26: 189-197.

[59] Bebawi F F, Cooper A P, Brodie G I, Madigan B A, Vitelli J S, Worsley K J and Davis K M (2007) Effect of microwave radiation on seed mortality of rubber vine (*Cryptostegia grandiflora* R.Br.), parthenium (*Parthenium hysterophorous* L.) and bellyache bush (*Jatropha gossypiifolia* L.). Plant Protection Quarterly. 22: 136-142.

[60] Ark P A and Parry W (1940) Application of High-Frequency Electrostatic Fields in Agriculture. The Quarterly Review of Biology. 15: 172-191.

[61] Tran V N (1979) Effects of Microwave Energy on the Strophiole, Seed Coat and Germination of Acacia Seeds. Australian Journal of Plant Physiology. 6: 277-287.

[62] Tran V N and Cavanagh A K (1979) Effects of microwave energy on Acacia longifolia. Journal of Microwave Power. 14: 21-27.

[63] Davis F S, Wayland J R and Merkle M G (1971) Ultrahigh-Frequency Electromagnetic Fields for Weed Control: Phytotoxicity and Selectivity. Science. 173: 535-537.

[64] Davis F S, Wayland, J. R. and Merkle, M. G. (1973) Phytotoxicity of a UHF Electromagnetic Field. Nature. 241: 291-292.

[65] Wolf W W, Vaughn C R, Harris R and Loper G M (1993) Insect radar cross-section for aerial density measurement and target classification. Transactions of the American Society of Agricultural and Biological Engineers. 36: 949-954.

[66] Barker A V and Craker L E (1991) Inhibition of weed seed germination by microwaves. Agronomy Journal. 83: 302-305.

[67] Haller H E (2002) Microwave Energy Applicator. United States Patent 20020090268A1,

[68] Clark W J and Kissell C W (2003) System and Method for In Situ Soil Sterilization, Insect Extermination and Weed Killing. United States Patent 20030215354A1,

[69] Grigorov G R (2003) Method and System for Exterminating Pests, Weeds and Pathogens. United States Patent 20030037482A1,

[70] Nelson S O (1996) A review and assessment of microwave energy for soil treatment to control pests. Transactions of the ASAE. 39: 281-289.

[71] Brodie G, Botta C and Woodworth J (2007) Preliminary investigation into microwave soil pasteurization using wheat as a test species. Plant Protection Quarterly. 22: 72-75.

[72] Brodie G, Harris G, Pasma L, Travers A, Leyson D, Lancaster C and Woodworth J (2009) Microwave soil heating for controlling ryegrass seed germination. Transactions of the American Society of Agricultural and Biological Engineers. 52: 295-302.

[73] Brodie G, Hamilton S and Woodworth J (2007) An assessment of microwave soil pasteurization for killing seeds and weeds. Plant Protection Quarterly. 22: 143-149.

[74] Connor F R (1972) Antennas. London: Edward Arnold.

[75] Brodie G, Ryan C and Lancaster C (2012) Microwave technologies as part of an integrated weed management strategy: A Review. International Journal of Agronomy. 2012: 1-14.

[76] Brodie G, Pasma L, Bennett H, Harris G and Woodworth J (2007) Evaluation of microwave soil pasteurization for controlling germination of perennial ryegrass (*Lolium perenne*) seeds. Plant Protection Quarterly. 22: 150-154.

[77] Brodie G, Ryan C and Lancaster C (2012) The effect of microwave radiation on Paddy Melon (Cucumis myriocarpus). International Journal of Agronomy. 2012: 1-10.

[78] Bigu-Del-Blanco J, Bristow J M and Romero-Sierra C (1977) Effects of low-level microwave radiation on germination and growth rate in corn seeds. Proceedings of the IEEE. 65: 1086-1088.

[79] Helsel Z R (1992) Energy and alternatives for fertilizer and pesticide use. In: Fluck R C editor. Energy in Farm Production. New York: Elsevier. pp. 177-201

[80] Hülsbergen K J, Feil B, Biermann S, Rathke G W, Kalk W D and Diepenbrock W (2001) A method of energy balancing in crop production and its application in a long-term fertilizer trial. Agriculture, Ecosystems & Environment. 86: 303-321.

[81] Mari G R and Chengying J (2007) Energy Analysis of various tillage and fertilizer treatments on corn production. American-Eurasian Journal of Agricultural and Environmental Science. 2: 486-497.

[82] Mari G R and Changying J (2007) Energy analysis of various tillage and fertilizer treatments on corn production. American-Eurasian Journal of Agricultural and Environmental Science. 2: 486-497.

[83] Langner H-R, Böttger H and Schmidt H (2006) A Special Vegetation Index for the Weed Detection in Sensor Based Precision Agriculture. Environmental Monitoring and Assessment. 117: 505-518.

[84] Heap I M (1997) The occurrence of herbicide-resistant weeds worldwide. Pesticide Science. 51: 235-243.

[85] Hill J M and Marchant T R (1996) Modelling Microwave Heating. Applied Mathematical Modeling. 20: 3-15.

[86] Vriezinga C A (1996) Thermal runaway and bistability in microwave heated isothermal slabs. Journal of Applied Physics. 79: 1779 -1783.

[87] Vriezinga C A (1999) Thermal profiles and thermal runaway in microwave heated slabs. Journal of Applied Physics. 85: 3774 -3779.

[88] Vriezinga C A, Sanchez-Pedreno S and Grasman J (2002) Thermal runaway in microwave heating: a mathematical analysis. Applied Mathematical Modelling. 26: 1029 -1038.

[89] Ulaby F T and El-Rayes M A (1987) Microwave Dielectric Spectrum of Vegetation - Part II: Dual-Dispersion Model. IEEE Transactions on Geoscience and Remote Sensing. GE-25: 550-557.

[90] El-Rayes M A and Ulaby F T (1987) Microwave Dielectric Spectrum of Vegetation-Part I: Experimental Observations. Geoscience and Remote Sensing, IEEE Transactions on. GE-25: 541-549.

[91] Zielonka P, Gierlik G, Matejak M and Dolowy K (1997) The comparison of experimental and theoretical temperature distribution during microwave wood heating. Holz als Roh- und Werkstoff. 55: 395-398.

[92] Jerby E, Dikhtyar V, Aktushev O and Grosglick U (2002) The microwave drill. Science. 298: 587-589.

[93] Jerby E, Aktushev, O., and Dikhtyar, V. (2005) Theoretical analysis of the microwave-drill near-field localized heating effect. Journal of applied physics. 97: 034909-1 - 034909-7.

[94] Torgovnikov G and Vinden P (2010) Microwave Wood Modification Technology and Its Applications. Forest Products Journal. 60: 173–182.

[95] Vinden P, Torgovnikov G and Hann J (2010) Microwave modification of Radiata pine railway sleepers for preservative treatment. European Journal of Wood and Wood Products. 1-9.

[96] Moriwaki S, Machida M, Tatsumoto H, Kuga M and Ogura T (2006) A study on thermal runaway of poly(vinyl chloride) by microwave irradiation. Journal of Analytical and Applied Pyrolysis. 76: 238 -242.

[97] Diprose M F, Benson F A and Willis A J (1984) The Effect of Externally Applied Electrostatic Fields, Microwave Radiation and Electric Currents on Plants and Other Organisms, with Special Reference to Weed Control. Botanical Review. 50: 171-223.

[98] Bird S, Brown W and Rowe J (2001) Safe and Effective Grain Feeding for Horses. Barton, ACT, Australia: Rural Industries Research and Development Corporation

[99] Dong S, Long R, Zhang D, Hu Z and Pu X (2005) Effect of microwave treatment on chemical composition and in sacco digestibility of wheat straw in yak cow Asian-Australasian Journal of Animal Sciences. 18: 27-31.

[100] Sadeghi A A and Shawrang P (2006) Effects of microwave irradiation on ruminal degradability and in vitro digestibility of canola meal. Animal Feed Science and Technology. 127: 45-54.

[101] Sadeghi A A and Shawrang P (2006) Effects of microwave irradiation on ruminal protein and starch degradation of corn grain. Animal Feed Science and Technology. 127: 113-123.

[102] Sadeghi A A, Nikkhaha A and Shawrang P (2005) Effects of microwave irradiation on ruminal degradation and in vitro digestibility of soya-bean meal. Animal Science. 80: 369-375.

[103] Brodie G, Rath C, Devanny M, Reeve J, Lancaster C, Harris G, Chaplin S and Laird C (2010) The effect of microwave treatment on lucerne fodder. Animal Production Science. 50: 124–129.

[104] McLeod M N and Minson D J (1978) The accuracy of the pepsin-cellulase technique for estimating the dry matter digestibility in vivo of grasses and legumes. Animal Feed Science and Technology. 3: 277-287.

[105] McLeod M N and Minson D J (1980) A note on Onozuka 3S cellulase as a replacement for Onozuka SS (P1500) cellulase when estimating forage digestibility in vitro. Animal Feed Science and Technology. 5: 347-350.

[106] McCleary B V, Gibson T S and Mugford D C (1997) Measurement of total starch in cereal products by amyloglucosidase-alpha-amylase method: Collaborative study. Journal of Aoac International. 80: 571-579.

[107] Chen S S and Spiro M (1994) Study of microwave extraction of essential oil constituents from plant materials. Journal of Microwave Power and Electromagnetic Energy. 29: 231-241.

[108] Saoud A A, Yunus R M and Aziz R A (2006) Yield study for extracted tea leaves essential oil using microwave-assisted process. American Journal of Chemical Engineering. 6: 22-27.

[109] Chemat S, Aït-Amar H, Lagha A and Esveld D C (2005) Microwave-assisted extraction kinetics of terpenes from caraway seeds. Chemical Engineering and Processing. 44: 1320-1326.

[110] Miletic P, Grujic R and Marjanovic-Balaban Z (2009) The application of microwaves in essential oil hydro-distillation processes. Chemical Industry and Chemical Engineering Quarterly. 15: 37-39.

[111] Brodie G, Jacob M V, Sheehan M, Yin L, Cushion M and Harris G (2011) Microwave modification of sugar cane to enhance juice extraction during milling. Journal of Microwave Power and Electromagnetic Energy. 45: 178-187.

[112] Vinden P and Torgovnikov G (1996) A method for increasing the permeability of wood. University of Melbourne Patent PO 0850/96,

[113] Vinden P and Torgovnikov G (2000) The physical manipulation of wood properties using microwave. In Proceedings of International Conference of IUFRO. pp. 240-247

[114] Torgovnikov G and Vinden P (2005) New microwave technology and equipment for wood modification. In: Schulz R L and Folz D C editors. Microwave and Radio

Frequency Applications: Proceedings of the Fourth World Congress on Microwave and Radio Frequency Applications. Arnold MD: The Microwave Working Group. pp. 91-98

[115] Fernandez Y, Arenillas A and Menendez J A (2011) Microwave heating applied to pyrolysis. In: Grundas S editor. Advances in Induction and Microwave Heating of Mineral and Organic Materials. Rijeka, Croatia: InTech. pp. 723-752

[116] Dominguez A, Menendez J A, Inguanzo M and Pis J J (2005) Investigations into the characteristics of oils produced from microwave pyrolysis of sewage sludge. Fuel Processing Technology. 86: 1007 -1020.

[117] Tsubaki S and Azuma J-i (2011) Application of microwave technology for utilization of recalcitrant biomass. In: Grundas S editor. Advances in Induction and Microwave Heating of Mineral and Organic Materials. Rijeka, Croatia: InTech. pp. 697-722

Chapter 3

Microwave Applications in Thermal Food Processing

Mohamed S. Shaheen, Khaled F. El-Massry,
Ahmed H. El-Ghorab and Faqir M. Anjum

Additional information is available at the end of the chapter

1. Introduction

In this chapter an overview of microwave heating as one method of thermal food processing is presented. Due to the limited space, this overview cannot be complete; instead some important theoretical information and also examples of practical uses at home and in industry are shown. This chapter provides a starting point, and the interested reader is directed to the references, where more information about the special themes discussed in this chapter can be found (Dehne, 1999). Additional to the references in the text the interested reader is also referred to two bibliographies that cover more or less all the published work on microwaves (W.H.O, 2012).

2. History

The development of dielectric heating applications in food industry started in the radio frequency range in the 1930s (Püschner, 1966). The desired energy transfer rate enhancement led to an increased frequency: the microwaves. The first patent, describing an industrial conveyor belt microwave system was issued in 1952 (Spencer, 1952), however its first application started 10 years later. This was caused by the need for high power microwave sources to be developed. The first major applications were finish drying of potato chips, pre-cooking of poultry and bacon, tempering of frozen food and drying of pasta (Decareau, 1985). Whereas the first applications were only temporarily successful, since the quality enhancement due to the microwave process could quickly be achieved by a more economic improvement of the conventional technique, the other techniques survived and are still successful in industrial application.

3. Uses, advantages and disadvantages of microwave heating applications

Today's uses range from these well known applications over pasteurization and sterilization to combined processes like microwave vacuum drying. The rather slow spread of food industrial microwave applications has a number of reasons: there is the conservatism of the food industry (Decareau, 1985) and its relatively low research budget. Linked to this, there are difficulties in moderating the problems of microwave heating applications. One of the main problems is that, in order to get good results, they need a high input of engineering intelligence.

Different from conventional heating systems, where satisfactory results can be achieved easily by intuition, good microwave application results often do need a lot of knowledge or experience to understand and moderate effects like uneven heating or the thermal runaway. Another disadvantage of microwave heating as opposed to conventional heating is the need for electrical energy, which is its most expensive form. Nevertheless, microwave heating has a number of quantitative and qualitative advantages over conventional heating techniques that make its adoption a serious proposition. One main advantage is the place where the heat is generated, namely the product itself. Because of this, the effect of small heat conductivities or heat transfer coefficients does not play such an important role. Therefore, larger pieces can be heated in a shorter time and with a more even temperature distribution. These advantages often yield an increased production.

4. Microwave applications

4.1. Application in food industries

Due to the very large number of microwave ovens in households, the food related industry not only uses microwaves for processing but also develops products and product properties especially for microwave heating. This way of product enhancement is called product engineering or formulation.

4.1.1. Baking and cooking

Detailed references to the baking process of bread, cakes, pastry etc. by the help of microwaves on industrial scale can be found. An enhanced throughput is achieved by an acceleration of the baking where the additional space needs for microwave power generators are negligible. Microwaves in baking are used in combination with conventional or infrared surface baking; this avoids the problem of the lack of crust formation and surface browning. An advantage of the combined process is the possible use of European soft wheat with high alpha-amylase and low protein content.

In contrast to conventional baking microwave heating inactivates this enzyme fast enough (due to a fast and uniform temperature rise in the whole product) to prevent the starch from extensive breakdown, and develops sufficient CO2 and steam to produce a highly porous (Decareau, 1986). One difficulty to be overcome was a microwavable baking pan, which is sufficiently heat resistant and not too expensive for commercial use. By 1982 patents had been issued overcoming this problem by using metal baking pans in microwave ovens (Schiffmann et al., 1981 and Schiffmann, 1982).

The main use of microwaves in the baking industry today is the microwave finishing, when the low heat conductivity lead to considerable higher baking times in the conventional process. A different process that also can be accelerated by application of microwave heating is (pre-) cooking. It has been established for (pre)cooking of poultry (Helmar et al., 2007), meat patties and bacon. Microwaves are the main energy source, to render the fat and coagulate the proteins by an increased temperature. In the same time the surface water is removed by a convective air flow. Another advantage of this technique is the valuable by-product namely rendered fat of high quality, which is used as food flavoring (Schiffmann, 1986).

4.1.2. Thawing and tempering

Thawing and tempering have received much less attention in the literature than most other food processing operations. In commercial practice there are relatively few controlled thawing systems. Frozen meat, fish, vegetables, fruit, butter and juice concentrate are common raw materials for many food-manufacturing operations. Frozen meat, as supplied to the industry, ranges in size and shape from complete hindquarters of beef to small breasts of lamb and poultry portions, although the majority of the material is `boned-out' and packed in boxes approximately 15 cm thick weighing between 20 and 40 kg. Fish is normally in plate frozen slabs; fruit and vegetables in boxes, bags or tubs; and juice in large barrels. Few processes can handle the frozen material and it is usually either thawed or tempered before further processing.

Thawing is usually regarded as complete when all the material has reached 0 0C and no free ice is present. This is the minimum temperature at which the meat can be boned or other products cut or separated by hand. Lower temperatures (e.g. -5 to -2 ^{0}C) are acceptable for product that is destined for mechanical chopping, but such material is `tempered' rather than thawed. The two processes should not be confused because tempering only constitutes the initial phase of a complete thawing process. Thawing is often considered as simply the reversal of the freezing process.

However, inherent in thawing is a major problem that does not occur in the freezing operation. The majority of the bacteria that cause spoilage or food poisoning are found on the surfaces of food. During the freezing operation, surface temperatures are reduced rapidly and bacterial multiplication is severely limited, with bacteria becoming completely dormant below -10 $^{0C.}$ In the thawing operation these same surface areas are the first to rise in temperature and bacterial multiplication can recommence. On large objects subjected to long uncontrolled thawing cycles, surface spoilage can occur before the centre regions have fully thawed.

Conventional thawing and tempering systems supply heat to the surface and then rely on conduction to transfer that heat into the centre of the product. A few, including microwave, use electromagnetic radiation to generate heat within the food. In selecting a thawing or tempering system for industrial use a balance must be struck between thawing time, appearance and bacteriological condition of the product, processing problems such as effluent disposal, and the capital and operating costs of the respective systems. Of these factors, thawing time is the principal criterion that often governs selection of the system. Appearance, bacteriological condition and weight loss are important if the material is to be sold in the thawed condition but are less so if it is for processing. The main detrimental effect of freezing and thawing meat is the large increase in the amount of proteinaceous fluid (drip) released on final cutting, yet the influence of thawing rate on drip production is not clear.

James and James (2002) reported that studies have shown that there was no significant effect of thawing rate on the volume of drip in beef or pork. Several authors concluded that fast thawing rates would produce increased drip, while others showed the opposite. Thawing times from -8 to 0 °C of less than 1 minute or greater than 2000 minutes led to increased drip loss (James et al., 2002). The results are therefore conflicting and provide no useful design data for optimizing a thawing system. With fish, fruit and vegetables ice formation during freezing breaks up cell structure and fluids are reduced during thawing. In microwave tempering processes the heating uniformity and the control of the end temperature are very important, since a localized melting would be coupled to a thermal runaway effect.

4.1.3. Drying

The benefits of microwave drying we should first have a quick look at the much more conventional method of air drying. As shown in Fig.1, a typical drying curve of a foodstuff can be subdivided into three phases. The first period is one of constant drying rate per unit of surface area. During this period the surface is kept wet by the constant capillary-driven flow of water from within the particle. The factors that determine and limit the rate of drying in the so-called `constant rate period' all describe the state of the air: temperature and relative humidity as well as air velocity (Erle, 2000).

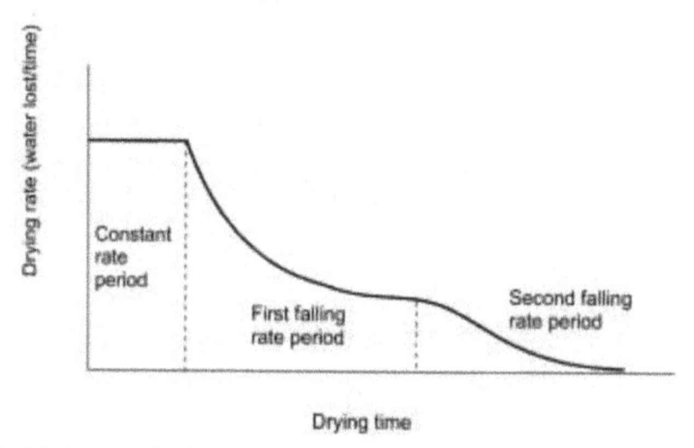

Figure 1. Typical drying curve for air drying

In drying the main cause for the application of microwaves is the acceleration of the processes, which are (without using microwaves) limited by low thermal conductivities, especially in products of low moisture content. Correspondingly sensorial and nutritional damage caused by long drying times or high surface temperatures can be prevented. The possible avoidance of case hardening, due to more homogeneous drying without large moisture gradients is another advantage. Two cases of microwave drying are possible, drying at atmospheric pressure and that with applied vacuum conditions.

Combined microwave-air-dryers are more widespread in the food industry, and can be classified into a serial or a parallel combination of the both methods. Applied examples for a serial hot air and microwave dehydration are pasta drying and the production of dried onions (Metaxas et al., 1983) whereas only intermittently successful in the 1960s and 1970s was the finish drying of potato chips. The combination of microwave and vacuum drying also has a certain potential. Microwave assisted freeze drying is well studied, but no commercial industrial application can be found, due to high costs and a small market for freeze dried food products (Knutson et al., 1987). Microwave vacuum drying with pressures above the triple point of water has more commercial potential has microwave vacuum drying with pressures above the triple point of water.

Microwave energy overcomes the problem of very high heat transfer and conduction resistances, leading to higher drying rates. These high drying rates correspond also to lower shrinkage and to the retention of water insoluble as shown in Figure 4. In parsley, for example, most of essential oils are present as a separate phase with high boiling temperature. For fast drying conditions (high microwave energy input) only the small amount of volatile essential oils that is dissolved is lost, whereas there is not enough time to resolve the remaining oil in the separated phase (Erle, 2000).

In contrast the retention of water soluble aromas, as in apples, is not as advantageous, since the microwave energy generates many vapour bubbles, so that the volatile aromas have a large surface to evaporate. Nevertheless, the low pressures limit the product temperatures to lower values, as long as a certain amount of free water is present and this helps to retain temperature sensitive substances like vitamins, colours etc. So, in some cases the high quality of the products could make also this relative expensive process economical.

Microwave vacuum dehydration is used for the concentration or even powder production of fruit juices and drying of grains in short times without germination .Newly and successfully applied is the combination of pre-air-drying, intermittent microwave vacuum drying (called puffing) and post-air-drying. It is predominantly used to produce dried fruits and vegetables, with improved rehydration properties (Räuber, 2000). After the form is stabilized by case hardening due to conventional air-drying, the microwave vacuum process opens the cell structures (puffing) due to the fast vapourization of water and an open pore structure is generated. The subsequent post-drying reduces the water content to the required value.

4.1.4. Quality of microwave-dried food products

In general, the quality is somewhere between air-dried and freeze-dried products. The reduction of drying times can be quite beneficial for the colour and the aroma. Venkatesh and Raghavan (2004) dried rosemary in a household microwave oven with good aroma retention. Krokida and Maroulis (1999) measured colour and porosity of microwave-dried apples, bananas, and carrots. Khraisheh et al. (2004) compared air-dried and microwave-dried potatoes and found a reduction of shrinkage and improved rehydration for the latter. Venkatesh et al (2004) reported on chicken products, seafood, and vegetables of good quality. He used air at 10±20 ^{0}C to cool the product during microwave drying. Quality can often be improved further by the use of vacuum. This reduces thermal as well as oxidative stress during processing.

For instance, Yongsawatdigul and Gunasekaran (1996) showed that colour and texture of microwave-vacuum-dried cranberries were better than those of air-dried samples. If we look specifically at the retention of aroma, it becomes necessary to distinguish between two basic cases. In most foods the aroma molecules are present in very small amounts, so that they are likely to be dissolved in the water phase. In this situation, the volatility of the aroma molecule in water is essential.

Considering the fact that we perceive aroma -as opposed to taste - with our noses, it is quite clear that aroma molecules are normally volatile; otherwise they would stay in the food during eating and not contribute to the aroma. In other words, if there is an interface between a water phase (i.e. a food) and a gas phase, the aroma molecules tend to choose the gas phase. In air drying, the surface where the aroma molecules can escape is mainly the outer surface of the particles. This is also where the water molecules evaporate. So the surface of the food particle will be depleted of aroma, but the losses cannot be higher than those that come with the capillary water flow from within. As a result, the losses of water and aroma are coupled.

5. Microwave drying applied in the food industry

Microwave drying is not common in the food industry. There are many reasons for its limited use: the technical problems described above were not well- understood in the past. This has led to some failures, which have surely discouraged other potential users. Schiffmann (2001) has listed a number of formerly successful applications that have been discontinued. Among these are the finish drying of potato chips, pasta drying, snack drying, and the finish drying of biscuits and crackers. It is apparently not always the microwave process itself but rather changes in the circumstances of production that make competing technologies more successful.

In spite of these difficulties, there are some current applications. Schiffmann (2001) mentions cereal cooking and drying with a production rate of nearly 1 ton/h. Pasta drying with microwaves is carried out in Italy. Microwave- vacuum drying is being used for meat extract and, at least for a number of years, for the production of a powder made from orange juice concentrate.

The combination of air drying and microwave-vacuum puffing is being used in Germany and Poland for fruits and vegetables. As the food industry does not disclose all its production processes, we cannot expect this list to be complete. Hauri (1989) has provided values for the necessary investment and the specific energy requirements of five different drying methods (Table 1). Based on the same throughput, the investment needed for microwave-vacuum drying is rather high, while the energy figures are more favorable than for air drying.

Types of drying process	Specific energy demand kwh/kg	Specific investment costs for equal throughput
Air band drying	1.9	100%
Spry drying	1.6	120%
Vacuum contact drying	1.3	150%
Microwave Vacuum drying	1.5	190%
Freez drying	2.0	230%

Table 1. Comparison of five different drying method

6. Pasteurization and sterilization

Studies of microwave assisted pasteurization and sterilization have been motivated by the fast and effective microwave heating of many foods containing water or salts. A detailed review can be found in (Rosenberg et al., 1987). Although, physically non-thermal effects on molecules are very improbable, early works seemed to show just these effects. But in most cases the results claimed could not be reproduced, or they lacked an exact temperature distribution determination. The improbability of non-thermal effects becomes clear, when the quantum energy of photons of microwaves, of a thermal radiator and the energy of molecular bonds are compared. The quantum energy of a photon of $f = 2.45$ GHz is defined by $E = hf \approx 1*10-5$ eV, the typical energy of a photon radiated from a body of 25°C ≈ 298 K equals $E = kT \approx 0.26$ eV and the energy of molecule bonds are in the eV-range.

Since the collection of energy with time for bound electrons are forbidden by quantum mechanics, only multi-photon processes, which are very unlikely, could yield chemical changes. Recently Lishchuk also showed that even a deviation of the energy distribution of water molecules from the conventional Boltzmann distribution cannot be proved (Lishchuk et al., 2001).

More thinkable is the induction of voltages and currents within living cell material, where eventual consequences are still in discussion (Sienkiewicz, 1998). Due to the unquestioned thermal effects of microwaves, they can be used for pasteurization and sterilization. Studied applications of microwave pasteurization or sterilization cover pre-packed food like yoghurt or pouch-packed meals as well as continuous pasteurization of fluids like milk (Helmar et al., 2007). Due to the corresponding product properties either conveyor belt systems or continuous resonator systems are invented.

The possibly high and nearly homogeneous heating rates, also in solid foods (heat generation within the food) and the corresponding short process times, which helps preserving a very high quality yield advantages of microwave compared to conventional techniques. The crucial point in both processes is the control and the knowledge of the lowest temperatures within the product, where the destruction of microorganisms has the slowest rate. Due to the difficult measurement or calculation of temperature profiles it is still very seldom industrially used.

7. Blanching using microwave processing

Blanching is an important step in the industrial processing of fruits and vegetables. It consists of a thermal process that can be performed by immersing vegetables in hot water (88-99 ^{0}C, the most common method), hot and boiling solutions containing acids and/or salts, steam, or microwaves. Blanching is carried out before freezing, frying, drying and canning. The main purpose of this process is to inactivate the enzyme systems that may cause color, flavor and textural changes, such as peroxidase, polyphenol-oxidase, lipoxygenase and pectin enzymes. The efficiency of the blanching process is usually based on the inactivation of one of the heat resistant enzymes: peroxidase or polyphenoloxidase.

Blanching has additional benefits, such as the cleansing of the product, the decreasing of the initial microbial load, exhausting gas from the plant tissue, and the preheating before processing. A moderate heating process such as blanching may also release carotenoids and make them more extractable and bioavailable (Arroqui et al., 2002).

However, this operation has also some inconvenient effects such as losses in product quality (texture and turgor), environmental impact, and energy costs. Leaching and degradation of nutritive components, such as sugars, minerals and vitamins, may occur when blanching with water or steam. The blanching process should assure enzyme inactivation while minimizing the negative effects, taking into account the interdependence of every aspect (Arroqui et al., 2002).

The use of microwaves for food processing has increased through the last decades. Some of the advantages compared with conventional heating methods include speed of operation, energy savings, precise process controls and faster start-up and shut-down times (Kidmose and Martens, 1999). Microwave blanching of fruits and vegetables is still limited. Some of the advantages compared with conventional heating methods include speed of operation and no additional water required. Hence there is a lower leaching of vitamins and other soluble nutrients, and the generation of waste water is eliminated or greatly reduced.

8. Applications of microwave blanching foods

Blanching with hot water after the microwave treatment compensates for any lack of heating uniformity that may have taken place, and also prevents desiccation or shriveling of delicate vegetables. And while microwave blanching alone provides a fresh vegetable flavor, the combination with initial water or steam blanching provides an economic advantage. This is because low-cost hot water or steam power is used to first partially raise the temperature, while microwave power, which costs more, does the more difficult task of internally blanching the food product.

A still further advantage is that microwave blanching enables a finish blanching of the center sections more quickly and without being affected by thick or non-uniform sections. Uniformity is also more rapidly accomplished in microwave ovens of the continuous tunnel types in contrast to the customary non-uniformity in institutional or domestic ovens (Smith and Williams, 1971).

The spraying of cold water at the end of the blanching process allows a better nutrient retention than the immersion of the food in cold water. Sub-atmospheric pressure, when applied to the steam blanching process, reduces the amount of oxygen and therefore results in a lower degradation of vegetable pigments and nutrients. Pressurized steam reduces blanching time. Optimal conditions of time, temperature, vapor pressure and microwave power depend on the particular vegetable that is being processed and must be empirically determined.

The knowledge of precise microwave power per weight of food that is needed to inactivate a particular enzyme should be sufficient to achieve a successful blanching and to avoid adverse effects. When the process temperature is not adequate, the enzymatic deteriorative action may prevail or even increase in some cases. Figure 2 shows the activity of mushroom polyphenol oxidase in a phosphate buffer 0.05M solution. The samples were previously treated in a microwave oven at specific times, using different potency levels: high, medium and low, which correspond to 770, 560 and 240 watts, respectively.

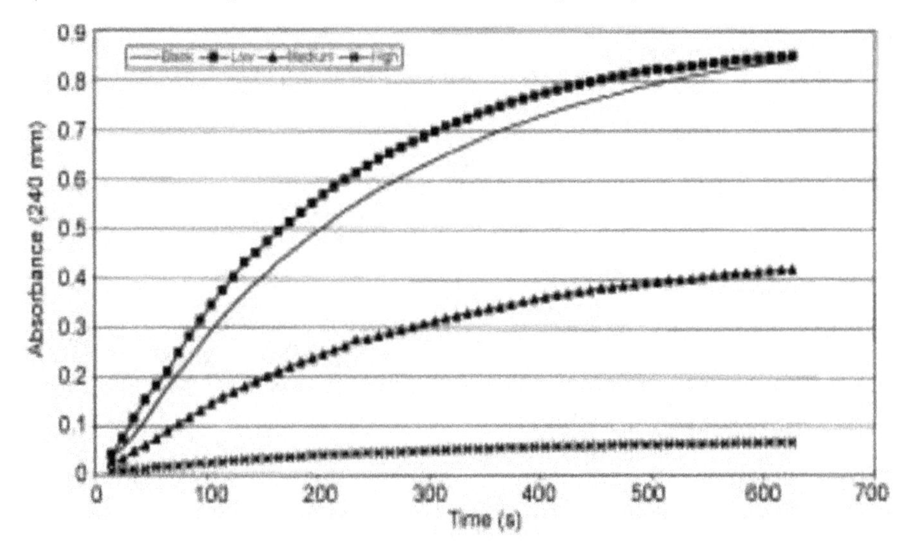

Figure 2. Mushroom tyrosinase as affected by microwaves.

9. Advantages of microwave blanching

Microwave heating involves conversion of electromagnetic energy into heat by selective absorption and dissipation. Microwave heating is attractive for heating of foods due to its origin within the material, fast temperature rise, controllable heat deposition, and easy clean-up. The very high frequencies used in microwave heating allow for rapid energy transfers and, thus, high rates of heating. These rates are a main advantage of this technique. Also, because microwaves penetrate the sample, heating is accomplished in the interior of the food. When heating rapidly, the quality of fruits and vegetables such as flavor, texture, color and vitamin content is better kept (Dorantes-Alvarez et al., 2000). However, rapid heating can also lead to problems of non-uniform heating when excessively high energy transfer rates are used (Ohlsson, 2000). It has been observed that microwave processing of chicken, beef, bacon, trout, and peanut oil does not change the fatty acid composition of these products, nor produces trans- isomers (Helmar et al., 2007).

10. Development of unique-single systems for microwave blanching

The most likely future for microwave food processing is in the continued development of unique single systems that overcome the limitations discussed previously. Compared to the development of traditional blanching systems, it is still a challenge to design appropriate equipment for microwave blanching.

This is due mainly to the following factors:

- Better control of the process is required due to the shorter heating times that microwave heating requires.
- The temperature distribution within the food product is affected by additional factors. A better distribution can be achieved by the use of standing and hold times at the end of the process. More research is needed in order to develop a method that would assure better repeatability of the process and equilibration of temperatures. The last objective can also be helped by a careful control of the food composition (Anantheswaran and Ramaswamy, 2001). Since the heating migration in microwave processing occurs from the initial and hottest locations in the interior of the food, it is difficult to locate and assess the cold point, as in traditional thermal methods. Therefore, the use of specific software to calculate the parameters of the process will help to achieve a higher efficiency (RodrõÂguez et al., 2003).

In the near future, it is expected that researchers interested in this matter will discover more specific effects that may be advantageous in the processing of food by microwave blanching. This would give an additional value to food products and would overcome the cost of microwave energy for this particular application.

11. Waste treatment under microwave irradiation

Many industrial activities involve the creation and subsequent disposal of waste, which represents a noticeable cost in terms of money and pollution. Moreover, sometimes waste materials are hazardous as well, i.e. materials containing asbestos or byproducts of nuclear plant. In this case, regulatory procedures are particularly restrictive, to guarantee the safety of the operators, and the choice of an inertization process becomes a compromise between safety issues, energetic evaluations and economical aspects. Thus, the waste treatment has to be evaluated nation of the final product.

The disposal of waste materials is now becoming a very serious problem, since in recent years the great increase of their production was not matched by a corresponding rise in the number of authorized dumps. Moreover, the existing regulation does not always allow all kind of waste material to be recycled, especially if harmful or hazardous materials are involved (Oda et al., 1992). But considering the present year production of wastes like ashes, or the wide spread presence on the territory of asbestos containing materials, it seems impossible to handle this environmental issue only by disposal in dumps. To face this situation, it is necessary to study and develop alternative ways to treat and re-use the components of waste materials, for instance converting them in secondary raw materials and, if possible, restoring them to accomplish the task they were initially meant for. Waste, even if originated by the same manufacturing process, and thus belonging to the same category (i.e. ashes, nuclear waste, asbestos containing materials, etc.), can be regarded as a multi-component material having a wide range of compositions, and usually it is the presence of only some of these components that makes all the mixture a product to be disposed of. Thus, a process allowing selective treatment of the "unwanted" portion of the waste, and to do this volumetrically, could represent an enormous advantage in terms of time and money, especially as far as materials presenting low thermal conductivity are concerned (Marucci et al., 2000). Microwaves can be an interesting candidate to fulfill the need for this kind of processes, and this is particularly true if the matrix of the waste materials exhibits dielectric properties significantly different from those of the unwanted components.

12. Safety of food processed in microwave for consumers

The food processed by this novel technology is safe for consumption. "Because the microwave energy is changed to heat as soon as it is absorbed by the food, it cannot make the food radioactive or contaminated (O.S.H.A, 2012). When the microwave energy is turned off and the food is removed from the oven, there is no residual radiation remaining in the food. In this regard, a microwave oven is much like an electric light that stops glowing when it is turned off (Gallawa, 2005).

13. Summary and outlook

Microwave ovens are commonplace in households and are established there as devices of everyday use. Their primary function is still the reheating of previously cooked or prepared meals. The relatively new combination of microwaves with other (e.g. conventional, infrared or air jet) heating systems should enhance their potential for a complete cooking device, that could replace conventional ovens. Unfortunately, in industry the distribution of microwave processes is still far away from such high numbers. Only a relatively low number of microwave applications can be found in actual industrial production, compared with their indisputable high potential. These successful microwave applications range over a great spectrum of all thermal food processes. The most prominent advantages of microwave heating are the reachable acceleration and time savings and the possible volume instead of surface heating. Reasons mentioned for the failure of industrial microwave applications range from high energy costs, which have to be counterbalanced by higher product qualities, over the conservatism of the food industry and relatively low research budgets, to the lack of microwave engineering knowledge and of complete microwave heating models and their calculation facilities. The latter disadvantage has been partly overcome by the exponentially growing calculating power which makes it possible to compute more and more realistic models by numerical methods. Very important for the task of realistic calculations is the determination of dielectric properties of food substances by experiments and theoretical approaches. Nevertheless in order to estimate results of microwave heating applications and to check roughly the numerical results, knowledge of simple solutions of the one-dimensional wave propagation like the exponentially damped wave is of practical (and also educational) relevance. But still the best test for numerical calculations is experiments, which yield the real temperature distributions within the product, which is really important especially in pasteurization and sterilization applications. While more conventional temperature probe systems, like fibre optic probes, liquid crystal foils or infrared photographs only give a kind of incomplete information about the temperature distribution within the whole sample, probably magnetic resonance imaging has the potential to give very useful information about the heating patterns. Hopefully, this together with the enormous calculation and modeling power will give the microwave technique an additional boost to become more widespread in industrial food production.

The breakthrough of microwave technology in the food industry due to its high potential has been predicted many times before, but it has been delayed every time up to now. That is why we are cautious in predicting the future of microwaves in industrial use. However, we think that the potential of microwave technology in the food industry is far from being exhausted.

Author details

Mohamed S. Shaheen, Khaled F. El–Massry* and Ahmed H. El–Ghorab
Flavour and Aroma Department, National Research Center, Egypt

Faqir M. Anjum
National Science& Technology (NIFSAT), Agriculture University, Faisalabad, Pakistan

* Corresponding Author

14. References

Anantheswaran, R. C. and Ramaswamy, H. S. (2001). `Bacterial destruction and enzyme inactivation during microwave heating', in Datta A K and Anantheswaran, R. C., Handbook of Microwave Technology for Food Applications, Marcel Dekker, New York, 191-210.

Arroqui, C., Rumsey, T. R., Lopez, A., and Virseda, P. (2002). `Losses by diffusion of ascorbic acid during recycled water blanching of potato tissue', J Food Eng, 52, 25-30.

Decareau, R. V. (1985), Microwaves in the Food Processing Industry. Orlando, Academic Press Inc.

Decareau, R. V., (1986). Microwave Food Processing Equipment Throughout the World,Food Technology, June 1986, 99-105.

Dehne, L. I. (1999), 'Bibliography on Microwave Heating of Food', Bundesinstitut Fur gesundheitlichen Verbraucherschutz und Veterinarmedizin, 04/1999

Dorantes-Alvarez, L., Barbosa-Caanovas, G., and Gutiearrez-Loa Pez, G. (2000). `Blanching of fruits and vegetables using microwaves', in Barbosa-CaÂnovas G and Gould G, Innovations of Food Processing, Technomic Publishing, Lancaster, PA, 149-162.

Erle, U. (2000).Untersuchungen zur Mikrowellen-Vakuumtrocknung von Lebensmitteln, Ph.D.Thesis, Universität Karlsruhe, Aachen, Shaker Verlag.

Gallawa, J.C. and Microtech Productions. Copyright © 1989-2005. http://www.gallawa.com/microtech/mwfaq.html.

Hauri, F.W. (1989). Vacuum band drying in food industry. Die ErnaÈhrungsindustrie, 10, 32-34.

Helmar, S. and Marc, R. (2007). The microwave processing of foods, Woodhead Publishing Limited and CRC Press, 20-312.

James, S.J. and James, C. (2002). Thawing and tempering. In Meat Refrigeration, Woodhead Publishing, Cambridge, 159-190.

Kidmose, U. and Martens, H. J. (1999). `Changes in texture, microstructure and nutritional quality of carrot slices during blanching and freezing', J Sci Food Agric, 79, 1747-1753.

Knutson, K. M.; Marth, E. H. and Wagner, M. K. (1987). 'Microwave Heating of Food', *Lebensmittel-Wissenschaft und -Technologie*, 20, 101-110,

Krokida, M. K., Kiranoudis, C. T., Maroulis, Z. B., and Marinos-Kouris, D. (2000), `Effect of pretreatment on color of dehydrated products', Drying Technology, 18, 1239-1250.

Krokida, M.K. and Maroulls, Z.B. (1999). Effect of microwave drying on some quality properties of dehydrated products. Drying Technology, 17(3),.449-466.

Lishchuk, S. V. and Fischer, J. (2001). 'Velocity Distribution of Water Molecules in Pores under Microwave Electric Field', International Journal of Thermal Sciences, Vol. 40(8), 717-723.

Marucci, G., Annibali, M., Carboni, G., Gherardi, G., Ragazzo ,G., Siligardi, C., Veronesi, P., Lusvarghi, L. and Rivasi, M.R. (2000) *"Characterization of microwave inertized asbestos containing materials"*, Materials Engineering Monograph, 115-125, Mucchi editore, Modena (IT).

Metaxas, A.C. (1996).Foundations of Electroheat- a Unified Approach, John Wiley and Sons, Chichester, UK,.

Oda, S. J. (1992). Microwave remediation of hazardous waste: a review, in Microwave processing of materials III, vol 269, 453-464, MRS,.

Ohlsson, T. (2000). `Minimal processing of foods with thermal methods', in Barbosa-CaÂnovas G and Gould G, Innovations of Food Processing, Technomic Publishing, Lancaster, PA.

Occupational Safety & Health Administration (OSHA), United States Department of Labor. Copyright ©2012. http://www.osha.gov/SLTC/radiofrequencyradiation/

Püschner, H. A., *Heating with Microwaves*, Berlin, Philips Technical Library, (1966)

Räuber, H. (1998). 'Instant-Gemüse aus dem östlichen Dreiländereck', *Gemüse*, 10'98,

Rodriaguez, J. J., Barbosa-Caa novas, G. V., Gutiearrez-Loa Pez. G. F., Dorantes-Alvarez, L., Won-Yeom, H., and Zhang, H. Q. (2003). `An update on some key alternative food processing technologies: microwave, pulsed electric field, high hydrostatic pressure, irradiation and ultrasound', in GutieÂrrez-LoÂpez G F and Barbosa-CaÂnovas G V, Food Science and Food Biotechnology, CRC Press, Washington D.C., 279-304.

Rosenberg, U. and Bög, W.(1987). 'Microwave Pasteurization, Sterilization, Blanching and Pest Control in the Food Industry', Food Technology, June 1987, 92-121,

Schiffmann, R. F.; Mirman, A. H.; Grillo, R. J., Microwave Proofing and Baking Bread Utilizing Metal Pans, U. S. Patent 4,271,203, (1981)

Schiffmann, R. F. (1982). Method of Baking Firm Bread, U. S. Patent 4,318,931.

Schiffmann, R. F. (1986). 'Food Product Development for Microwave Processing', Food Technology, June 1986, 94-98.

Schiffmann, R. F.; Mirman, A. H. and Grillo, R. J. (1981). Microwave Proofing and Baking Bread Utilizing Metal Pans, U. S. Patent 4,271-203.

Schiffmann, R.F. (2001). Microwave processes for the food industry. In: Datta, A.K. and Anantheswaran, R.C. (eds.), Handbook of Microwave Technology for Food Applications. New York: Marcel Dekker.

Sienkiewicz, Z. (1998). 'Biological Effects of Electromagnetic Fields', Power Engineering Journal, 12 3, 131-139.

Smith, F. J. and Williams, L. G. (1971). `Microwave blanching', United States Patent No.3,578,463, patented 11 May 1971, applicant: Cryodry Corporation, San Ramon, CA.

Spencer, P., Means for Treating Foodstuffs, U. S. Patent 2,605,383, 605, 383, (1952)

Venkatesh, M. S. and Raghavan, G .S V. (2004), `An overview of microwave processing and dielectric properties of agri-food materials', Biosys Eng, 88(1), 1-18.

Yongsawatdigul, J. and Gunasekaran, S. (1996). Microwave-vacuum drying of cranberries: Part II. Quality evaluation. Journal of Food Processing and Preservation, 20, 145-156.

World Health Organization. Copyright © 2012. http://www.who.int/peh-emf/about/WhatisEMF/en/index4.html.

Chapter 4

Effect of Microwave Heating on Flavour Generation and Food Processing

G.E. Ibrahim, A.H. El-Ghorab,
K.F. El-Massry and F. Osman

Additional information is available at the end of the chapter

1. Introduction

The flavour and colour of a food product have significant impacts on consumer acceptability. Two of the challenges with microwave food products are that it is often difficult to achieve the desired flavour that matches products prepared in a conventional oven or by frying and to get the browning that the consumer expects. There are reactions that occur in those processes that do not occur when foods are heated in the microwave oven and this is part of what contributes to the lack of flavour and colour development. In trying to solve the flavour issues, it is important to understand what flavours are as well as all of the attributes of a food product that lead to consumer liking. Solving the colour problem involves an understanding of the reactions that produce colour and finding ways to get those reactions to occur. The typical browning which occurs when foods are heated by conventional means produces not only the desired brown pigments but also produces a variety of desirable flavours.

Flavours and colours generated as a result of the Maillard reaction are of critical importance for the commercial success of microwave-processed foods. Recent interest in the microwave generation of Maillard flavours and colours was a response on the part of the food industry, based on the consumer demand for fast and convenient food products. The fundamental differences between microwave and conventional heating, the composition of the food matrix, and the design of microwave ovens all seem to play a role in the inability of microwave heating to propagate colou and flavourting Maillard reactions in food products. The increased sales of microwave ovens in the last decade, especially into the North American market, provided the food industry with the impetus for renewed interest in carrying out the Maillard reaction in microwaveable food products.

1.1. Flavour definition

Flavour is defined as the experience of the combined perception of compounds responsible for taste and aroma. The flavour of food is very important for its acceptability and a slight change in the odour of processed food may affect the overall quality of the product. Aromas come from low molecular weight organic compounds that can volatilize and be sensed in the nasal cavity. These compounds are not of one simple class of chemicals but rather are many different chemical types, including acids, esters, alcohols, ketones, pyrazines, thiazoles and terpenes as well as many others. The human body has a complex set of receptors that recognize both individual compounds as well as mixture of compounds to identify different flavours.

Character impact compounds are individual chemicals with a specific, recognizable aroma. Some examples are methyl anthranilate (concord grape), citral (lemon), cinnamic aldehyde (cinnamon), methyl salicylate (wintergreen) and diacetyl (butter). While these individual compounds have a characteristic odour, they do not make up the complete flavour of a product, whether naturally occurring such as in a concord grape or in a flavour added to a product. Many other compounds are also present which build the overall flavour profile. Considerable work has been done to identify the flavour compounds in different foods. There are over 170 compounds that have been identified that contribute to the flavour of a strawberry, while coffee and chocolate are much more complex with over 800 compounds identified.

1.2. Natural versus artificial flavours

In the United States, there is a legal definition of natural and artificial flavours.

The complete definitions are found in the Code of Federal Regulations (CFR) Title 21 101.22 (Code of Federal Regulations, 2008). Artificial flavours are defined in (a)(1) as follows:

The term artificial flavour or artificial flavouring means any substance, the function of which is to impart flavour, which is not derived from a spice, fruit or fruit juice, vegetable or vegetable juice, edible yeast, herb, bark, bud, root, leaf or similar plant material, meat, fish, poultry, eggs, dairy products, or fermentation products thereof. Artificial flavour includes the substances listed in Sec. 172.515(b) and 182.60 of this chapter except where these are derived from natural sources.

Natural flavours are defined in (a)(3) of Title 21 101.22, as follows:

The term natural flavour or natural flavouring means the essential oil, oleoresin, essence or extractive, protein hydrolysate, distillate, or any product of roasting, heating or enzymolysis, which contains the flavouring constituents derived from a spice, fruit or fruit juice, vegetable or vegetable juice, edible yeast, herb, bark, bud, root, leaf or similar plant material, meat, seafood, poultry, eggs, dairy products, or fermentation products thereof, whose significant function in food is flavouring rather than nutritional. Natural flavours include the natural essence or extractives obtained from plants listed in Sec. 182.10, 182.20,

182.40, and 182.50 and part 184 of this chapter, and the substances listed in Sec. 172.510 of this chapter.

The browning reaction will be discussed later and it is interesting to note that process flavours can be made using the browning reaction and are defined as natural since they are a product of roasting or heating.

Spices are also defined in (a)(2) of Title 21 101.22 as follows:

The term spice means any aromatic vegetable substance in the whole, broken, or ground form, except for those substances which have been traditionally regarded as foods, such as onions, garlic and celery; whose significant function in food is seasoning rather than nutritional; that is true to name; and from which no portion of any volatile oil or other flavouring principle has been removed.

The regulation goes on to list a number of individual spices. It should be noted that many materials that are considered artificial are identical to those in nature: it is simply how they were produced that determines whether they are natural or artificial. As an example, diacetyl is natural if it comes from milk or is produced by fermentation (as in wine and fermented dairy products) but is artificial if it is synthesized from other chemicals. All of the chemical properties are the same no matter where the individual chemical came from. The only time that a compound can never be natural is if it has never been found from any natural source. One compound, which gives a cotton candy type flavour, is ethyl maltol. This has never been found in nature so if a flavor contains this compound, it will be at least partially artificial.

In other countries, there are different definitions as to what is natural and artificial. In some countries, there is the concept of nature identical. It states that if a compound exists in nature, then it does not matter where it comes from, it would not be considered artificial. The flavour could not be called natural but has simply been referred to as flavour. It is important that the regulations for each country be checked to understand what is allowed and how flavours added to products should be labeled.

2. Sources of flavours

Flavouring materials come from a variety of sources. One of the main sources is plants. The flavour materials can be present in any part of a plant including the flower, leaf, stem or bark. To be used in food products, the materials are generally extracted from the plant material to provide an isolate that is just the flavour. There are different techniques that can be used for isolation including solvent extraction (often ethanol), steam distillation and supercritical fluid extraction. Dairy products and meats and seafood can also be sources of flavouring materials. Dairy products provide a good source of base material that can be modified by enzymes to create much more concentrated flavors than are present in the natural dairy product. The enzymes break down the fats and proteins present to yield higher concentrations of the flavour compounds that represent the flavors of these dairy products. Some of the components of dairy flavors are short chain fatty acids including butyric acid that are unpleasant at high concentrations but help to contribute to the characteristic flavour of dairy products.

Many flavours are produced by processing, primarily with the use of heat. The subject of browning will be covered in more detail later in this chapter but will be briefly addressed here. Flavours can be created by heating one or more reducing sugars with one or more amino acids for different times and at different temperatures. Very different flavours can be produced which can be added to foods as natural flavours.

Flavours can also be produced using biotechnology. This is an area that has been explored for years to determine ways to get plants or microorganisms to produce higher quantities of flavouring materials than they do naturally. While there has been limited success by some companies to produce individual flavour compounds through this process, it has not achieved wide commercial success.

3. Microwave versus conventional heating

The industrial and domestic use of microwaves has increased dramatically over the past few decades. While the use of large-scale microwave processes is increasing, recent improvements in the design of high-powered microwave ovens, reduced equipment manufacturing costs and trends in electrical energy costs offer a significant potential for developing new and improved industrial microwave processes.

Microwave heating is relatively fast compared to conventional heating since it does not depend on the slower diffusion process in the latter. This property initiated the initial investigation into carrying out chemical reactions under microwave irradiation (Giguere et al., 1986). In certain cases, chemical reactions were completed in a few seconds that otherwise would have taken hours. In addition to fast rates of heating, microwaves are also more selective and components can be heated selectively in a reaction mixture compared to conventional heating. This property has been used to enhance the extraction of essential oils from plants immersed in a microwave transparent solvent (Paré et al., 1991).

Microwaves are electromagnetic waves within a frequency band of 300MHz to 300 GHz. In the electromagnetic spectrum **(Fig. 1)** they are embedded between the radio frequency range at lower frequencies and infrared and visible light at higher frequencies. Thus, microwaves belong to the non-ionising radiations.

Superheating of solvents is another phenomena that accompanies microwave heating and helps accelerate chemical reactions. Superheating refers to the increase in temperature of liquids above their boiling points while they remain completely in the liquid phase. For example, water boils under microwave heating at 105°C and acetonitrile (B.P. 82°C) at 120°C. A chemical reaction carried out in an open vessel in acetonitrile under microwave irradiation will be accelerated by 14 times relative to conventional heating, assuming the reaction rate doubles for every 10°C rise in temperature (Peterson, 1993). When chemical reactions are carried out in closed containers under microwave irradiation, the maximum temperature attainable is not limited to the temperature of the heating medium, as in conventional heating, but depends only on the microwave power applied and the rate at which the sample can lose heat. The extreme high temperatures attained in a closed container during microwave heating can generate extreme high pressures (especially if the reaction produces gaseous products), which can alter equilibrium product distribution according to Le Chatelier's principle (Peterson, 1993).

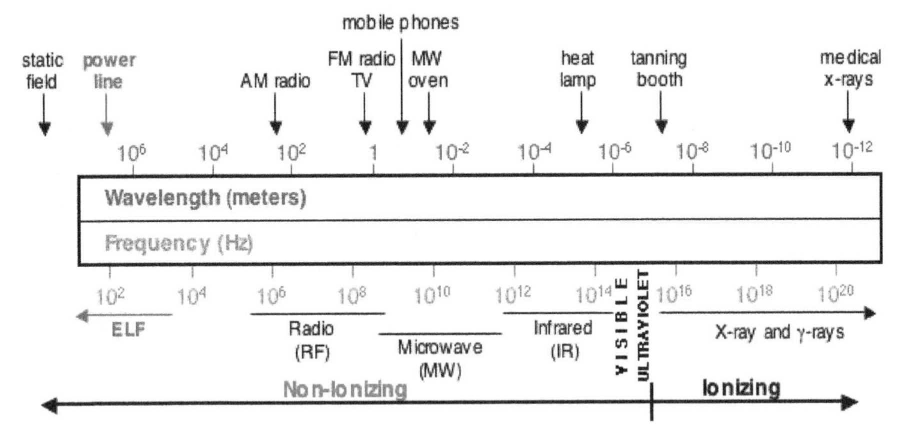

Figure 1. Electromagnetic spectrum. Additionally, the two most commonly used microwave frequency bands (at 915MHz and 2450 MHz) are sketched.

There are several major factors that impact the flavour quality of microwave food products. They primarily stem from the fact that in a conventional oven, the product is surrounded by hot air which heats the product from the outside and also dries the surface. In microwave heating, the entire product is heated at the same time but the heating may not be uniform (van Eijk, 1994). In drying the surface, it helps to reduce the rate at which volatile flavour molecules can move from inside the product to the surface and evaporate. It in a sense forms a crust that is more difficult for the flavour molecules to move through. In microwave heated products, the surface stays moist and cooler, which readily allow flavour compounds to be carried out of the food as steam is lost.

The surface of the product will also get to a higher temperature in a conventional oven. This enhances the rate of the browning reaction on the surface as this reaction goes more rapidly under lower moisture and higher temperature conditions. The browning reaction provides not only the desirable brown colour but also produces a large number of flavour compounds. In conventionally heated products, the added flavour is retained better and a large number of flavours are produced on the browned surface of the product. In products where browning is not expected, this is not an issue. If a product is simply to be reheated, the microwave does an excellent job as you are not relying on it to produce flavour. One additional factor that influences flavour development in products heated in the microwave is that they are in the oven for a much shorter period of time than those cooked in a conventional oven. The browning reaction takes time to develop and the product is not heated long enough for this reaction to proceed to the point where brown pigments and flavour compounds are produced. It should be noted that there are a wide variety of products where the time and temperature of heating do not create an issue for flavour development. High moisture products that are going to be reheated work very well. While some flavour will be lost during the heating process, it does not vary significantly from conventional reheating. Vegetables, with their own inherent flavor, can easily be steamed in the microwave oven.

The sensory properties of vacuum-microwave-dried and air-dried carrot slices, which were water blanched initially. The vacuum-microwave-dried carrot slices received the higher ratings for texture, odour and overall acceptability as compared to the air-dried carrot slices.

The retention of volatile components responsible for flavour was more in hot air microwave drying compared to conventional hot air drying alone. The flavour strength of garlic dried by hot air alone is 3.27 mg/g dry matter whereas the flavour strength of the garlic dried by microwave drying is 4.06 mg/g dry matter. Effect of microwave drying on the shelf life and sensory attributes (appearance, colour, odour and overall quality) of coriander (*Coriander sativum*), mint (*Mentha spicata*), fenugreek (*Trigonella foenum-graceum*), amaranth (*Amaranthus sp.*) and shepu (*Peucedanum graveolens*).

Amaranth had similar scores for fresh and dried ones; however, there was significant decrease for the sensory attributes of other greens. They concluded that microwave drying was highly suitable for amaranth, moderately suitable for shepu and fenugreek and less suitable for coriander and mint. Wheat samples were evaluated and the sensory characteristics of grain were assessed by the panel of 10 members. The sample produced a burnt or roasted odour when exposed for a long exposure time (180 s) but there was no significant difference in the grain odour when long exposure times were avoided.

Due to high temperature and long drying time, volatile compounds are vapourised and are lost with water vapour, resulting in significant loss of characteristic flavour in dried products. Case-hardening is a common problem in dried fruits due to rapid drying. As drying proceeds, the rate of water evaporation is faster than the rate of water movement to the product surface, hence making the outer skin dry

At air dryer temperatures, volatile flavour compounds are lost, structural changes such as case hardening may inhibit later rehydration, and extended drying times allow chemical and enzymatic reactions to degrade vitamins, flavour and colour compounds microwave dried frozen berries had a higher rehydration ratio. Microwave (MW) drying generated three unique flavou compounds (2-butanone, 2-methyl butanal, and 3-methyl butanal) while freeze-dried berries lost several, including the typical blueberry aroma, 1,8-cineole. Compared with hot-air dried berries, MW-dried cranberries have better colour, softer texture and similar

The advantages of MW blanching (MB) over conventional heat blanching methods (water or steam) include in-depth heating without a temperature gradient, and rapid inactivation of enzyme complexes that cause quality degradation coupled with minimal leaching of vitamins, flavours, pigments, carbohydrates, and other water-soluble components. No differences existed for flavour of green beans and mustard greens due to blanching method. In beans and mustard greens, steam blanching produced a texture equal to MB vegetables but chlorophyll degradation was greater. Cooking time of chicken breasts increased with decreasing power level, but cooking losses were not affected. Both sensory and instrumental tenderness (Instron compression) were best at 60% power level, while juiciness, mealiness and flavour were unaffected by power level. Convectional MW-cooked chicken was more tender, juicy and acceptable than MW-cooked chicken, avour intensity was similar. Thiamin retention ranged from 77% in conventionally cooked chicken breasts to 98% in MW-cooked chicken legs.

4. Microwave food process design

In designing microwave food processes and packaging, various factors that affect microwave heating of foods should be taken into consideration if the effect of uneven heating associated with the use of microwaves is to be kept under control. These factors fall into two broad categories. The first one is thermo-physical properties of the food. The second is factors associated with the dielectric characteristics of food and the field intensity distributions provided by various microwave energy applicators and heating systems.

4.1. Physical factors

The thermo-physical factors that require serious consideration in the design of microwave food processes and packaging systems are:

4.1.1. Size and shape of food

The physical size and shape of foods affect the temperature distribution within the food. This results from the fact that the intensity of the wave decreases with depth as it penetrates the food. If the physical dimensions of the food are greater than twice the penetration depth of the wave, portions of the food nearer the surface can have very high temperatures while the mid-portions are still cold. On the other hand, if the dimensions of the food are much lower than the penetration depth of the wave, the center temperature can be far higher than the temperature at the surface. This situation normally results in "the focusing effect," which results from the combined intensity of the wave (in three space dimensions) being higher at the inner portions than the outer portions of the product.

Some shapes reflect more microwaves than others. In addition, some shapes prevent increasing amounts of the waves from leaving the material by reflecting them back into the interior. For most spherical and cylindrical foods, wave focusing occurs for product diameters between 20 and 60 mm. In rectangular foods, focusing causes the overheating of corners. Thus in package design, sharp corners are avoided and tube-shaped pans have been suggested (Giese 1992). Moreover, in foods with corners, packages are designed using metals or aluminum foils to reflect microwave energy away from corners and thus selectively heat some portions more.

4.1.2. Surface area

In microwave heating, the product temperature rises above its ambient temperature due to volumetric heating. Higher product surface area therefore results in higher surface heat loss rate and more rapid surface cooling. During microwave heating, the highest temperature is not at the surface of the product (despite the higher intensity of power absorbed there) but somewhere in the interior.

4.1.3. Specific heat

How much a food product will heat given a specific amount of energy depends on its heat capacity. The implication of this for microwave heating is that different food products heated together have different temperature histories. To control this, some microwave food packages are sealed tight to allow heat transfer between hotter and colder foods, thus giving similar temperature history for different foods in the same package.

5. Microwave heating and the dielectric properties of foods

Microwave energy is transported as an electromagnetic wave in certain frequency bands in the range between about 0.3 GHz and 300 GHz. When microwaves impinge on a dielectric material, part of the energy is transmitted, part reflected and part absorbed by the material where it is dissipated as heat. Heating is due to `molecular friction' of permanent dipoles within the material as they try to reorient themselves with the oscillating (electrical) field of the incident wave. The power generated in a material is proportional to the frequency of the source, the dielectric loss of the material, and the square of the field strength within it. A material is subjected to microwave energy in a device known as an applicator or cavity. Considering all these features, it is possible to identify those candidate materials and processes that can use microwave heating effectively and understand microwave ingredient interaction mechanisms. Only after such a step is taken can microwave heating be exploited fully in terms of its unique characteristics, which include the facts that no contact is required between the energy source and the target and that heating is volumetric, rapid and highly specific in nature.

International convention dictates that microwave ovens (and other industrial, scientific and medical microwave applications) operate at specific frequencies, the most favoured being 2.45 GHz. At this frequency the electric field swings the orientation of water molecules 109 times every second, creating an intense heat that can escalate as quickly as 10 ^{0}C per second (Lew et al., 2002). Water being the predominant component of biological materials, its content directly influences heating. However, there are minor contributions from a host of other factors (Schiffmann, 1986): heating is accelerated by ionic effects (mostly salt content) and specific heat of the composite material (Decareau, 1992). Specific heat is an important property in the thermal behaviour of a food subjected to microwaves. Produce with low specific heat may heat very rapidly, and even faster than water of the same weight. Oil heats faster than water due to its much lower specific heat (Schiffmann, 1986). Hence for oily materials, the influence of specific heat becomes the determining factor in microwave heating, owing to the low specific heat of oils, often less than half that of water (Ohlsson, 1983).

5.1. Microwave interactions with dielectric properties

When an oscillating electrical field is applied to a polar dielectric, the dipoles within the material attempt to align themselves (polarize) with the field. The rate of change of polarization represents a displacement current in the dielectric and the product of this and the applied field gives the power generated as heat. Averaged over a cycle, the power `lost' in the material (i.e. dissipated as heat) depends on the phase angle between the applied field and the polarization. For most dielectrics the lag depends on the flexibility of the molecules that house the dipoles, and the randomization effect of temperature.

6. Flavour generation *via* Maillard reaction

The Maillard reaction is incredibly complex. For instance, a simple example such as the reaction of glucose with ammonia gives evidence, using simple methods, of the formation of more than 15 compounds and the reaction of glucose with glycine gives more than 24. Using HPLC and TLC on solvent-soluble material only [0.1% (w/w) of reactants], about 100 components are detectable as reaction products of xylose and glycine (Hodge, 1953).In order to understand something so complex, it is necessary to draw up a simplified scheme of the reactions involved. This has been done most successfully by Hodge in 1953 (**Fig. 2**).

The discussion here is based on this.

Hodge subdivides the Maillard reaction as follows:

i. Initial stage: products colourless, without absorption in the ultraviolet (about 280 nm).
   Reaction A: Sugar–amine condensation
   Reaction B: Amadori rearrangement
ii. Intermediate stage: products colourless or yellow, with strong absorption in the ultraviolet.
   Reaction C: Sugar dehydration
   Reaction D: Sugar fragmentation
   Reaction E: Amino acid degradation (Strecker degradation)
iii. Final stage: products highly coloured.
   Reaction F: Aldol condensation
   Reaction G: Aldehyde–amine condensation and formation of heterocyclic nitrogen compounds.

It is worth noting that Mauron (1981) calls the three stages Early, Advanced, and Final Maillard reactions, respectively. The way these reactions fit together is outlined in **Fig.2**. The final products of nonenzymic browning are called melanoidins to distinguish them from the melanins produced by enzymic browning. Theoretically, the distinction is clear; however, in practice, it is very difficult to classify the dark brown products formed in foods, since they tend to be very complex mixtures and are chemically relatively intractable. Reaction H has been inserted into **Fig.2.** It represents the much more recently discovered free-radical breakdown of Maillard intermediates. Oxygen plays an essential role in enzymic browning, but is not essential for nonenzymic browning. It may help in fact, for example, in the formation of reductones, such as dehydroascorbic acid, but it may also hinder the progress of the reaction, for example, in oxidising 2-oxopropanal to 2-oxopropanoic acid.

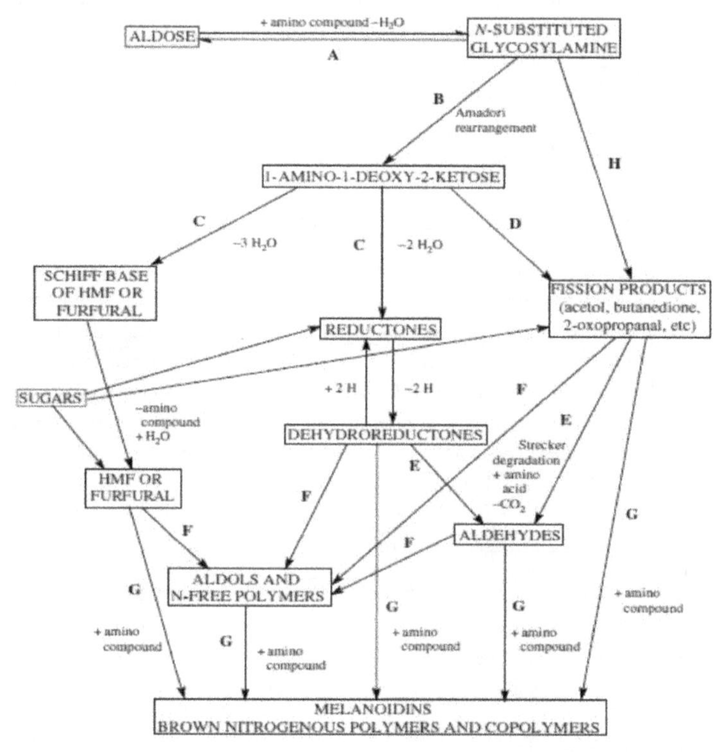

Figure 2. Maillard reaction.

In relation to the flavours produced on exposure to microwave radiation, Yaylayan et al.(1994) examined many combinations of sugars and amino acids, grouping the latter into aliphatic, hydroxylated, aromatic, secondary, basic, amide, acid, and sulfur-containing ones. The odours observed were grouped into eight and they have been assigned to the above groups, as far as possible, below:

Caramel (1); Meaty (4); Nutty (9); Meaty + vegetable Fragrant (6); Baked potato (5) and Baked (3).

Shibamoto and Yeo (1994) have compared microwave (700 W, high setting, 15 min) and thermal treatment (reflux, 100 °C, 40 h) for a glucose–cysteine system. The conditions used were determined by the onset of browning and aroma formation. The two sets of conditions gave samples with similar popcorn and nutty flavours, but the microwaved samples also gave pungent, raw, and burnt aromas, absent from the conventionally heated ones. The sample prepared conventionally at pH 9 contained much higher amounts of methylpyrazine and 2,6-dimethylpyrazine, whereas the microwaved one gave a much higher amount of 4,5-dimethyloxazole and was the only one to produce 2,3-dihydro-3,5-dihydroxy-6-methyl-4H-pyran-4-one. Such data, to some extent, explain the differences in acceptability of the two types of heating. For browning to occur in microwaving, a minimum of 10% moisture is required. Surprisingly, microwaving at pH 2 gave about twice the absorption at 420 nm than at pH 9 (about 1 AU), the absorption for pH 5 and 7 samples being nearly 0 (<0.1 AU).

As the Maillard reaction is a series of chemical transformations (Yaylayan, 1997) factors that influence a chemical reaction also affect the Maillard reaction. In general, the rates of chemical reactions depend primarily on temperature, pressure, time, and concentration of reactants. High temperature, pressure, and superheating of reaction solvent associated with microwave irradiation can accelerate simple or single-step chemical reactions such as esterification, hydrolysis and cyclization reactions (Richard et al., 1988; Bose et al., 1994). If the microwave heating is performed under a closed system, then the rate of the microwave reaction accelerates up to 1000 times. However, the time factor plays a crucial role in influencing the product distribution of more complex reactions when carried out under microwave heating. The influence on competitive and consecutive reactions is an important consequence of fast rate of heating under microwave irradiation that is especially pertinent to the propagation of Maillard reaction.

7. Solvent-mediated Maillard reactions: Model systems

Given the fact that the Maillard reaction is a complex series of consecutive and competitive reactions, product distribution and intensity of browning will be most affected by microwave irradiation relative to conventional heating. Generally, the final outcome of a Maillard reaction (colour, volatile aroma compounds, and nonvolatile products) depends on temperature, water content, pH, and heating time. Thus, any variation in the reaction parameters will affect the profile of the end products, and hence the perceived aroma and colour. Although simple chemical reactions are fast under microwave irradiation, multistep reactions can remain incomplete or they do not proceed to the same extent as under conventional heating. They produce mixtures that contain the same products (Yaylayan et al., 1994) but with altered distribution patterns. The flavour perception is sensitive to such variations in relative concentrations of different components, especially the character impact compounds, thus drastically changing the sensory properties.

There are few reports in the literature on the microwave-assisted generation of Maillard products using precursors or intermediates. Preparative scale microwave-assisted synthesis (Shui et al., 1990) of Amadori products from D-glucose and amino acids is feasible but has not been reported. However, Barbiroli et al. (1978) observed 70–75% conversion of added glucose/leucine into Amadori compounds with a corresponding decrease in the amount of added amino acid in a bread mix when microwaved for 3 min. Steinke et al. (1989) generated Strecker aldehydes from an aqueous solution of an amino acid and 2,3-butadione (diacetyl) in sealed vials microwaved for 4 min or heated in a water bath for 60 min at the same temperature. Significantly higher concentrations of aldehydes were measured in the microwave heated samples. The effect of electrolytes and pH on the formation of Maillard products during microwave irradiation of aqueous model systems has been studied.

The addition of different salts (Yeo and Shibamoto, 1991a) such as sodium chloride, calcium chloride, and sodium sulfate increased both the intensity of browning and the concentration of flavour compounds. The total volatiles generated from a glucose/cysteine model system (Yeo, and Shibamoto, 1991b) under microwave irradiation has been found to increase with pH. It seems that increasing the pH and concentration of electrolytes enhances the rate of Maillard reactions under microwave irradiation.

Attempts have been made to compare the chemical composition and yields of volatiles in microwaved and conventionally heated Maillard model systems. However, this type of comparison can be misleading due to the variations in the time-temperature exposure of the two systems under study. In most cases, the temperature of the microwave system is not monitored and time of irradiation is chosen arbitrarily. In order to compare the yields of two systems undergoing the same reaction at different times and temperatures, knowledge of kinetic parameters is required to ascertain whether there are differences in the two processes. Alternatively, the intensity of brown colour formation can be used as an indication that the two systems have undergone equivalent time-temperature exposure. Yaylayan et al. (1994) mimicked actual cooking and surface drying of foods by subjecting the same aqueous sugar/amino acid mixtures to microwave irradiation (640 W) and to conventional heating in an open system, until all the water was evaporated and the residue was dark brown. In order to ensure that both treatments produced the same extent of Maillard reaction for comparison purposes, the conventional heating time was adjusted such that after similar dilutions, both samples had the same spectrophotometric absorption at 460 nm. On the average, 1 min of microwave heating time produced the same browning extent as 12 min of conventional heating time. With such treatment, no significant qualitative changes were observed in the composition of both samples, as identified by GC/MS. Parliment (1993) studied, in sealed vials, the products of the Maillard reaction between glucose and proline formed under microwave (600 W, preheated conventionally for 3 min and irradiated for 45 s) and conventionally heated systems (150°C for 15 min). Qualitatively both systems produced similar compounds but in the microwave system N-heterocyclic compounds were present in smaller amounts.

Inhibition of pyrazine formation by natural antioxidants and the foods containing them was measured in a microwaved glucose/glycine model system. Inhibition of lipid oxidation by the same materials was assayed in both bulk and emulsion systems.

Pyrazines were determined by solid-phase micro extraction followed by GC. Lipid oxidation volatiles were assayed by polyamide fluorescence produced by either a bulk oil display or a hematin- or 2,2′-azobis-(2-amidino-propane) dihydrochloride-accelerated lecithin or fish oil emulsion. It was shown that (i) the inhibition of pyrazine formation depends on high concentrations of water-soluble antioxidants; (ii) such antioxidants occur naturally in some foods and are usually polyphenols; (iii) during pyrazine inhibition, oxidized polyphenols show enhanced nonfluorescing browning similar to enzymic browning products; (iv) monophenols, which structurally cannot form quinone polymers on oxidation, inhibit pyrazines with less browning; (v) during the final pyrazine-forming phase of the Maillard reaction, polyphenolics and reducing agents such as glutathione and ascorbic acid are partially consumed with some nutritional loss; (vi) fruit powders of grape seed, grape skin, and red wine are highly pyrazineinhibitory, steeped blueberry strongly so, but plum purees are moderately pro-pyrazine, and freeze-dried vegetables strongly pro-pyrazine; and (vii) black and green tea infusions are highly inhibitory, whereas spices have mixed effects.

7.1. Interaction of microwave with the food components

The major food components - water, carbohydrates, lipids, proteins and salts (minerals) - interact differently with MW. Because the primary mechanisms of MW heating are dipole rotation and ion acceleration, MW interactions with foods depend heavily on salt and moisture content. Water selectively absorbs the energy (Mudgett, 1990). In intermediate and high moisture products, the water, not the solids, absorbs the MW energy (Mudgett, 1989; Karel, 1975). However, because of their high heat capacity, they tend to heat unevenly. In drier products, the dissolved salts are concentrated (in the remaining water); if the solids exceed saturation level and precipitate, their ionic conductivities are limited. However, the solids themselves do absorb energy (marshmallow ignition: Mudgett, 1989). Low moisture products generally heat more evenly due to their low heat capacity (Schiffman, 1986).

Alcohols and the hydroxyl groups on sugars and carbohydrates are capable of forming hydrogen bonds and undergo dipolar rotation in an electric field. Low levels of alcohols or sugars in solution in foods have little effect on the interaction of MW with water and dissolved ions. At higher concentrations (jellies, candies), sugars can alter the frequency response of water with MW (Mudgett, 1989).

Proteins have ionizable surface regions that may bind water (or salts), giving rise to various effects associated with free surface charge. Lipids, other than the charged carboxyl groups of the fatty acids, which are usually unavailable due to their participation in the ester linkages of triglycerides, are hydrophobic and interact little with MW if water is present. MW do appear to interact with lipids (and colloidal solids) in low moisture foods as evidenced by energy absorption that cannot be accounted for by either free water or ion activity.

The interaction of microwave energy and food products causes internal heat generation. The rapidly alternating electromagnetic field produces intraparticle collisions in the material, and the translational kinetic energy is converted into heat. For many food products the heating is uneven; the outer layers heat most rapidly, depending on the depth of penetration of the energy, and the heat is subsequently conducted into the body of the food. Current research is concerned with achieving uniform heating, especially in relation to pasteurization and sterilization of foods, where non-uniform heating could result in a failure to achieve a safe process. For materials that are electrical conductors – *e.g.* metals, which have a very low resistivity – microwave energy is not absorbed but reflected, and heating does not occur. Short-circuiting may result unless the container is suitably designed and positioned. Metallic containers and trays can effectively improve the uniformity of heating (George 1993; George and Campbell 1994). Currently most packages are made of plastic materials which are transparent to microwave energy. The amount of heat generated in microwave heating depends upon the dielectric properties of the food and the loss factor (see below), which are affected by the food composition, the temperature and the frequency of the microwave energy. For tables of electrical properties of food, and discussion of their application Kent (1987), Ryynanen (1995), and Calay et al. (1995).

The main frequency bands used are 2450 and 896 MHz in Europe and 915 MHz in the USA. Greater penetration and more uniform heating are obtained at the longer wavelengths for food products with low loss factors. Datta and Liu (1992) have compared microwave and conventional heating of foods and concluded that microwave heating is not always the most effective method, especially for nutrient preservation. The effect depends on a variety of properties of the system.

Burfoot et al. (1988) examined the microwave pasteurization of prepared meals using a continuous tunnel device. The product was heated to 80–85°C for a few minutes, sufficient to inactivate vegetative pathogenic bacteria, *e.g.* Salmonella and Campylobacter, but not bacterial spores. The latter are controlled by storing the product below 10°C. This type of product is not shelf-stable at room temperature and a full sterilization process would be necessary with low acid products of this type to obtain a stable product. Microwave tunnels for this purpose would have to be pressurized to maintain the integrity of the package when sterilizing temperatures (121°C) had been achieved. A general-purpose plant known as Multitherm has been developed by AlfaStar Ab, Tumba, Sweden (Hallstrom et al. 1988). Burfoot et al. (1996) have modeled the pasteurization of simulated prepared meals in plastic trays with microwaves. Large differences between actual and predicted temperatures were found at some points. For measuring temperatures in microwave systems an invasive fibre-optic probe has been developed, which uses the change in color with temperature of a crystal situated at the end of a glass fibre. Fluoroptic probes are manufactured by Luxtron Corp., CA, USA.

Variations in electric fields, food constituents and the location of the food in a MW oven can lead to nonuniform heating, allowing for less-than-ideal interaction of food components and survival of microorganisms. A number of techniques to improve uniformity of MW heating, such as rotating and oscillating foods, providing an absorbing medium (water) around the product, cycling the power (pulsed power), and varying the frequency and phase, can improve the situation; however, dielectric properties of the food must be known in order to develop effective processes (Yang and Gunasekaran, 2001; Guan et al., 2004).

Using moisture, salt, and fat content, and temperature (<70 °C) at MW frequencies, Calay et al. (1995) developed polynomial equations to estimate dielectric properties of grains, fruits and vegetables, and meat products. However, they concluded that it was impossible to develop a generic composition based equation. This may be, in part, because as cooking temperature increases, the dielectric constant may increase while the loss factor and depth of penetration decrease (Zheng et al., 1998). The result is that changes in formulation usually require re-evaluation with regard to dielectric properties and behavior upon exposure to MW energy.

7.2. Maillard reaction interaction with food matrix during microwave irradiation

At the molecular level, the mechanism of heat generation in the microwave oven relies mainly on the interaction of the microwave radiation with dipoles/induced dipoles or with ions. Proteins and lipids do not significantly interact with microwave radiation in the presence of aqueous ions that selectively absorb the radiation. However, in the absence of water, lipids and colloidal solids are known to interact strongly with microwave radiation and the observed levels of energy absorption cannot be explained by the presence of free water and by ion activity (Pomeanz and Meloan, 1987).

Microwave radiation can also interact with alcohols, sugars, and polysachharides. Tightly bound water monolayers do not absorb energy due to hindered molecular rotations. Microwave interactions with a multicomponent system such as food can differ considerably from simple aqueous Maillard model systems, in that "matrix effects" can produce undesirable consequences. Since the core aqueous region of foods are the main sites of interaction with the microwaves, the interior vapor pressure generated as a result can actively force the vapor to the surface of the food, unlike in the conventional oven, where passive migration of water by capillary action to the surface, is diffusion controlled (Schiffmann, 1994a). The water-saturated food surfaces usually remain at relatively cool temperatures of the oven during cooking (40–60°C), thus preventing browning and crisping (Schiffmann, 1994b). Model studies have already indicated that there are no fundamental differences in the solution phase chemistry of Maillard reaction under microwave irradiation. However, the overall performance of food products under microwave irradiation implies the development of characteristic textural, color and aroma properties similar to that of conventional heating, which differs markedly from microwave heating due to fast rate of heating and "matrix effects."

Food products that rely heavily on Maillard flavours and colours, such as roasted and baked products, perform well in the conventional oven due to the following:

1. The high temperature of the air surrounding the product dehydrates the surface, producing a crust that protects the food from loss of moisture and important aroma volatiles. Dehydration steps are also crucial for the formation of color and aroma precursors by the Maillard reaction.
2. Long time exposure in the conventional oven ensures the completion of slow and/or multistep Maillard reactions responsible for browning and for the generation of specific aromas.

3. In the case of porous materials such as bread, the high temperature and relative low humidity of the air surrounding the product cause rapid heating of the surface of the food relative to the center, thus creating a temperature and a corresponding inward vapor pressure gradient that helps retain volatile aroma compounds inside the core.

In the microwave oven, the short time exposure and the lack of hot dry air (air being transparent to microwave irradiation) surrounding the surface of the food product not only prevents crusting but also promotes sogginess due to the condensation of the moisture. On the other hand, the rapid release of moisture and its evaporation from the center of the food causes the added and formed volatiles to be "steam distilled" at temperatures below their boiling points. Hence baked and roasted food products, which rely heavily on Maillard produced flavors, usually do not perform well in the microwave oven.

Schiffmann (1994a) summarized the different factors related to microwave ovens that affect aroma generation during cooking of food such as variation in the type of commercial ovens (power, cavity size, ...*etc.*) and its effect on the reproducibility of performance, speed of heating, oven temperature, and vapor pressure buildup inside the food. The short time required in the microwave oven to attain the same temperature as in the conventional oven not only retards the Maillard reaction but also prevents the establishment of thermal equilibrium throughout the food and uniform temperature distribution through conductive heat transfer. These hot and cold spots in the food product aggravate further the oven hot and cold zones created as a result of standing wave patterns. In addition, different dielectric loss factors (ϵ'') associated with different components in a multi component food product will also contribute to the uneven heating pattern inside the microwave oven. The combined effect of these phenomena is manifested in the excessive exposure of certain parts of food to heat and diminished exposure in others, leading to undesirable textural and flavour modifications such as charring, drying, excessive evaporation, hardening, and development of burnt or raw flavor and aroma notes. The extent of these undesirable modifications is dependent on the size, geometry, thickness, and the composition of the food product. Yeo and Shibamoto (1991c) reviewed the chemical composition of volatiles generated by microwave and conventionally heated food products. White cake batters were cooked to the same degree both in the microwave and the conventional oven. The volatiles released and sensory properties of both products were compared (Whorton and Reineccius, 1989). The number of volatiles detected and the amount of total pyrazines produced were found to be more in the conventionally baked sample. In addition, the microwave cake lacked the nutty, caramel, and browned flavors. In a similar study (Mac-Leod and Coppock, 1976) the number of volatiles generated from boiled beef cooked by microwave for 1 h, was found to be more than the number of volatiles generated by beef boiled conventionally, for the same length of time. When both systems were compared on the basis of "doneness," the microwave sample generated only one third the amount of volatiles detected in the conventional oven. The relative success of the microwave to achieve the Maillard effect of conventional heating may depend to a large degree on the type and composition of the food product.

8. Application of microwave technology in some food products

The use of microwave energy in food processing can be classified into six unit operations: (re)heating, baking and (pre)cooking, tempering, blanching, pasteurisation and sterilisation, and dehydration. Although their objectives differ, these aims are established by similar means: an increase in temperature. Nevertheless, for each special use (different from pure microwave heating), different advantages and disadvantages have to be taken into account. These are presented in the next sections together with some examples of real industrial applications.

Microwaves lend well to speeding up almost any drying processing which the liquid being evaporated is neither explosive nor flammable. The great advantage of microwave drying is speed, often allowing the drying of a material in 10% or less of the normal drying time. However, in no application, other than laboratory analytical drying systems, are microwaves used to dry a product alone. Always there is the use of additional heat—hot air, ambient or forced circulation; infrared; or some combination of these. In fact, microwave heating, properly applied, and usually represents a minor part of the total heat energy required for drying, the reason being cost.

Also, the following benefits encourage the application of microwave drying technology:

- The products obtain excellent rehydration properties because of the volumetric vaporisation of water, which is a constituent of most food stuffs. When evaporated simultaneously through the whole product piece volume, the water forms capillars inside the product pieces, which produce a high porosity and therefore allow an easy rehydration. In many cases gaining porosity is associated with an expansion of volume, so this process is called „puffing" sometimes. The instant properties are crucial for the application of dried food in the composition of ready meals and additives used in modern citchens.
- The same effect of gaining porosity enables the production of crispy fruit and vegetable snacks.
- In comparison to conventional air belt drying the application of microwaves under vacuum conditions allows fast volumetric drying of the product pieces at relatively low temperatures. So vitamins, taste, flavour and natural colours are conserved very well.
- In comparison to other advanced drying technologies (*i.e.* freeze drying) microwave vacuum drying is is more economic, as drying progress is much faster and thus allows a higher througput for the same plant dimensions.
- The combination of microwave vacuum drying or puffing and conventional airbelt pre-drying into a continuous production line is economically most advantageous.

8.1. Drying

Drying occurs when water vapor pressure differences between the food interior and exterior drive moisture transfer into the surrounding air. MW drying occurs by both dielectric and conventional heating. When above 50% moisture, as moisture content decreases, dielectric constant and loss factor decrease, especially at higher temperatures. Below 30% moisture content, MW penetration depth increases sharply (Feng et al., 2002). MW heating in a drying system may adversely affect product quality due to nonuniform temperature distribution and difficulty in controlling product final temperature at low moisture contents.

MW energy can improve quality of fried products. Potato chips can be fried then dried by MW and hot air (Decareau, 1985). MW finish drying to maintain the temperature below the Maillard browning point, of russet burbank potato slices containing <0.9% reducing sugar, allows production of chips of acceptable color and texture (Porter, 1971). Potatoes containing >0.9% reducing sugar must be removed from the oil at an intermediate moisture content >13% to obtain acceptable color of the MW-finished product. Oil content of MW-finished chips may be 90% that of conventional chips because the fat is absorbed at prefinish moisture levels. Osmotic dehydration prior to MW dehydration efficiently removes water from fruits and preserves volatile flavor compounds. Prothon et al. (2002) osmotically dried apple cubes in 50% (w/w) sucrose, then dried them in a MW-assisted drier. Osmotic dehydration reduced drying time required to reach 10% moisture, but also decreased drying rate and effective moisture diffusivity.

Osmotic pretreatment increased cell wall thickness and increased firmness frehydrated apple pieces, but reduced rehydration capacity. Drying is more efficient when strawberries and blueberries are pretreated with 2% ethyl oleate and 0.5% NaOH (osmotic drying: Venkatachalapathy and Raghavan, 1998, 1999). The osmotic dehydration step was necessary to produce MW-dried strawberries that had similar rehydration ratio, texture, color and sensory properties to freeze-dried berries. Dipping blueberries in 2.5% ethyl oleate and 0.2% NaOH followed by sucrose osmotic dehydration prior to MW drying treatment reduces drying (from >80% to 15% moisture) time to one-twentieth of that needed for tray drying (Feng et al., 1999). MW-dried frozen berries had a higher rehydration ratio. MW drying generated three unique flavour compounds (2-butanone, 2-methyl butanal, and 3-methyl butanal) while freeze-dried berries lost several, including the typical blueberry aroma, 1,8-cineole. Compared with hot-air dried berries, MW-dried cranberries have better color, softer texture and similar storage stability at room temperature (Yongsawatdigul and Gunasekaran, 1996). Vacuum permits water vaporization at a lower temperature and at a faster rate than at atmospheric pressure. Application of vacuum reduces the boiling point of water and the drying temperatures. Combining vacuum and MW drying (VMD) reduces or avoids the heat and rate limitations at atmospheric pressure (Durance and Wang, 2002). MW energy is an efficient mechanism of energy transfer through the vacuum and into the interior of the food. Drying time for carrots has been shown to be 30% less for a combination of VMD and hot air drying than that of a conventional hot air drying method (Baysal et al., 2002). No constant rate period existed and drying occurred mainly during the falling rate period. No differences occurred in dry matter content, bulk density or porosity; however, aw and color (L, a, b values) were higher and rehydration capacities were higher in carrots dried by the combination method. Fruit and vegetable variety can have significant effects on the VMD process.

After blanching potatoes prior to VMD to produce fat-free chips, Lefort et al. (2003) reported that yellow flesh cultivars had lower moisture content and higher specific gravity, starch content, and crispness scores than red flesh cultivars. The authors concluded that cultivars low in specific gravity and starch content produced chips with a crispy but less rigid texture, which are desirable characteristics for chips produced by VMD. Color was unaffected. A CaCl2 pretreatment prior to MW-assisted AD increases the hardness of rehydrated apples and potatoes (Arhne et al., 2003). Water loss rates are similar during drying at 50 ^{0}C, but at 70 ^{0}C rates in potatoes are slower Retention of volatiles makes VMD an attractive preservation method for herbs and spices. Parsley subjected to VMD is greener immediately and after 8 weeks than hot air-dried samples (Boehm et al., 2002). VMD preserved more than 90% of the essential oils compared to 30% by hot air drying and resulting in higher parsley-like and green-grassy aroma and less hay/straw-like off-flavor. MW drying of a variety of herbs, requiring 10 to 16 min, affected color, appearance, aroma and relative reconstitution capacity (RRC; Fathima et al., 2001). The RRC for dried coriander, mint, fenugreek, shepu and amaranthus was 10.3, 10.3, 31.7, 32.8, and 38.3 respectively. Herbs with the lowest RRC (mint, coriander), had the lowest scores for flavor and color scores, while dried amaranthus, with the highest RRC, had scores similar to that of the fresh herb. Storage (60 d) results in little change in sensory properties. Working with garlic, Sharma and Prasad (2001) reported that in comparison with hot air drying (70 ^{0}C) alone, VMD reduced drying time by 80-90% and dried garlic products had higher sensory quality scores. Yousif et al. (1999) found that VMD basil yielded 2.5 times the linalool and 1.5 times methylchavicol (the major volatiles) as air-dried samples. VMD basil had more volatiles than fresh basil due to chemical reactions during drying. AD basil was darker and less green. VMD samples had a higher rehydration rate, while the potential of the plant material to rehydrate was hindered in AD samples possibly due to maintenance of structural integrity of the cells.

Begum and Brewer (2001) studied the physical, chemical and sensory quality of snow peas blanched by boiling water, steam, microwave and microwave blanching in heat-sealable bags. No differences occurred in lightness L values. Boiling water-blanched peas were the least green, with low a values, whereas steam and microwave blanched in bag peas were the most green. Boiling water-blanched peas had the least b value whereas there was no significant difference in the b values for all other blanching treatments. Ascorbic acid, one of the most labile nutrients in vegetable is water soluble and sensitive to pH, light and heat and is affected by the naturally occurring enzyme ascorbic acid oxidase. Preservation of ascorbic acid in vegetables, particularly those that are good sources, is important in preserving food quality (Brewer and Begum, 2003). Ascorbic acid losses in fruits and vegetables are inevitable and all blanching treatments result in some reduced ascorbic acid losses. Boiling water blanching produced the lowest reduced ascorbic acid content in snow peas while all other blanching treatments resulted in 31–32 mg/100 g. At higher power and longer times actual reduced ascorbic acid content increased, but when adjusted for moisture losses, reduced ascorbic acid content decreased. Lane et al. (1985) studied the ascorbic acid content of four vegetables blanched by microwave and conventional methods (boiling water and steaming). Their results suggested that with the exception of steam-blanched purple hull peas, ascorbic acid retention was not affected by the blanching method. Drake et al. (1981) studied the influence of blanching method on the quality of selected vegetables. Water-and steam-blanched asparagus and green beans had similar ascorbic acid concentrations and both were superior in ascorbic acid to the microwave-blanched product. Microwave and steam-blanched green peas contained less ascorbic acid than water-blanched green peas.

8.2. Baking

Microwave baking has been the focus of much research and development since the 1950s, with variable success (Seyhun et al., 2003). All results point to the general rule that to achieve success, considerable product reformulation must be considered (Decareau, 1992). Baking with various emulsifiers, gums, starches, fat contents and enzymes has been widely investigated (Ozmutlu et al., 2001; Sumnu, 2001; Keskin et al., 2004). With adequate product formulation, microwave baking can offer good quality products with high convenience. Pillsbury® has a new line of frozen biscuits and dinner rolls specifically designed for the microwave to deliver warm, soft, ready-to-eat bread rolls in 25 s.

The browning reactions in baked products are the result of heating of reducing sugars with proteins or nitrogen-containing substances to form compounds like melanoidins, and start at around 160 ^{0}C (Matz, 1960). When sugars such as fructose, maltose and dextrose are heated to around 171 ^{0}C, molecules are combined to form coloured substances called caramels. A relatively low food surface and low surrounding temperatures in microwave baking do not enable the browning reactions to occur. Moreover, in microwave ovens, evaporated water molecules from the food system directly interact with cold air around the product and condense, which prevents browning and crisping reactions (Schiffmann, 1994). Dough products which are expected to be crisp and brown become soggy after baking. When heated for a longer period, they become dry and brittle but never brown. Brown surfaces achieved by Maillard reactions and caramelization of sugars are a result of high temperature accompanied by dehydration (Burea et al., 1987). In addition, time is necessary for completion of these browning reactions. The kinetic rate constant of browning reaction increases with increased temperature (Ibarz et al., 2000) and decreased moisture content (Moyano et al., 2002).

Good browning and flavour development were attained when no water was added to the system. The characteristic baked/roasted aroma produced decreased as moisture content increased. Flavour development was still apparent in the 5% moisture system (GC/MS analysis of flavour compounds) despite very little browning. Electrolytes (0-0.5M NaCl, $CaCl_2$, $FeCl_2$, or $NaSO_3$) enhanced both flavor production and browning intensity in an L-cysteine/D-glucose model system. NaCl promoted the development of the greatest amount of volatiles (seven times the control) and $FeCl_2$ the least (three Microwave processing, nutritional and sensory quality 93 times the control). NaCl produced the most browning while $FeCl_2$ produced the least. Browning treatments for breads were evaluated. When susceptors were used with MW baking, desired browning and hardness were obtained on the bottom surfaces of the breads but did not affect surface color significantly. Breads coated with the solution containing sodium bicarbonate (10.5%), glucose (31.6%) and glycine (5.3%) did not have the desired crust colour or hardness, while conventional browning at 200 ^{0}C achieved browning on top and bottom surfaces and crust formation on the bottom surface in 8 min.

MW-baked dough products are often of lower quality than conventionally baked products. Differences in heat and mass transfer patterns, insufficient starch gelatinization due to very short MW baking times, MW-induced changes in gluten, and rapid generation of gas and steam result in crustless products which are tougher and coarser and have less firm textures than conventionally baked products. They often have reduced height, gummy texture, hard crumb, and an undesirable moisture gradient along the vertical axis of the product. During baking, two simultaneous processes occur: (1) energy (heat) is transferred to the food and (2) this causes changes (starch gelatinization, protein denaturation) within and at the surface of the product. In conventional baking, the pattern of temperature rise in the interior differs substantially from that near its surface. During MW heating, the dough near the surface is heated instantaneously but heat must be transferred to the interior via conduction. This instantaneous surface heating promotes nearly instantaneous water evaporation as well. In addition, the cellular structure of dough makes it a poor heat conductor.

Flavours generated as a result of browning reactions are also absent in microwave baked products. The aroma profile of a microwave baked cake was shown to be similar to that of batter. Many of the nutty, brown and caramel-type aromas observed in the conventional cake were lacking in microwave baked cakes (Whorton and Reineccius, 1990). Individual flavour components are subjected to losses through distillation, flavour binding by starches and proteins and chemical degradation during microwave baking. Crust also provides a barrier against the loss of flavours (Eliasson and Larsson, 1993). Flavours can easily be released from microwave baked product due to the absence of crust.

Changing the food formulation to reduce compound volatility minimizes loss of flavour compounds during microwave baking. This can be done by adding an oil phase or increasing the oil content (Yaylayan and Roberts, 2001). Flavouring agents may be encapsulated to reduce the volatility of aroma compounds (Whorton and Reineccius, 1990). Unwanted flavours such as flour or egg-like flavours develop during microwave baking of cakes. Flavouring agents may be added to mask these unwanted flavours and obtain a similar flavour profile to conventionally baked cakes. Since products baked in microwave ovens have inferior quality, improving this quality represents a challenge to food technologists. Therefore, a thorough understanding of the effects of microwaves on the major ingredients in baked products such as starch and gluten will play an important role in improving the quality of these products.

8.3. Cooking

Microwave heating of food products is done in a relatively quick time period as compared to conventional oven cooking. The flavour of the final product can result from aroma generated during microwave cooking. It can also be already contained in the food, *e.g.*, in a precooked meal. In any case, the release phenomenon is the same and it is the timing that may be different. If the flavour is only produced at the end of heating, losses due to volatilization will be diminished considerably. In microwave heating of a precooked product, the volatile aroma compounds are integral to its final aroma, and losses can imbalance the aroma. Aroma is defined as the volatile aroma compounds that contribute both to the orthonasal (sniffing) and retronasal (eating) smell of a food. This section will explain the theory and give examples of how aroma compounds present in microwave foods can be lost during cooking.

The impact of microwave cooking on the formation of early Maillard products was investigated and compared with the effect of conventional cooking, using milk as a test system. Experiments were carried out at controlled temperatures of 80°C and 90°C, respectively, at holding times up to 420 min. Hydroxymethylfurfural (HMF) and lactulose, which are all established indicators to estimate heat damage, were determined. The concentrations of all the heating indicators increased with increasing heating time. For example in the 90°C test series the furosine values rose from 34 mg litre^{-1} (0.5 h) to 94 mg litre^{-1} (2 h holding time) in the milk heated by microwaves and from 35 mg litre^{-1} (0.5 h) to 96 mg litre^{-1} (2 h) in the conventionally heated milk. None of the reaction products showed significant differences as between the microwave heating and conventional cooking methods (Katz, 1994).

Flavour may be a problem in MW-cooked foods because flavour volatiles distill off, bind to proteins and other molecules or fail to develop at all. A number of methods have been developed to prevent or offset these flavor problems. Extraction process wherein substrates are mixed with MW-transparent solvent and exposed to MW which liberates target compounds from natural materials (*e.g.* spices). Selectivity can be varied by altering solvents/conditions. MW extraction in combination with liquid CO_2 can be used as an alternative to supercritical fluid extraction (decaffeination of coffee, defatting of cocoa powder). A process to generate desirable aromas when a food and/or package is subjected to MW radiation. The aroma-generating material, consisting of a sugar alone or in combination with an amino acid source, and an effective amount of a MW susceptible material for conductive heat transfer sufficient to catalyze the desired chemical reactions

Vegetables are often cooked to increase palatability and digestibility, ascorbic acid content of MW-cooked, frozen peas was lower, retention of chlorophyll and organic acids (lactic, succinic, malic, citric......*etc*) was higher for peas cooked without water. Effects were smaller for carrots. Those cooked without water had higher flavor scores and carotene retention than those cooked with water. While MW cooking of vegetables generally results in better nutrient retention, there is no one method that produces overall superior sensory characteristics when considering color, flavor, texture, and moistness.

Cooking starchy tubers gelatinizes the starch softening the texture. After a lag of 4 min, water loss during MW cooking of potatoes was rapid and linear. Starch gelatinization began at the surface and in the center, and then spread throughout the tuber cross-section after 1 min. Results suggest that the MW cooking process is divided into two phases: (1) the MW energy input raises the internal temperature to about 100 °C, then (2) water is vaporized at a constant temperature. Immersing potatoes in boiling water after the first phase prolonged cooking time compared to MW heating, suggesting that MW treatment affects texture by a mechanism independent of the thermal profile induced by cooking.

9. Conclusion

This chapter reviewed the flavours and colors appropriate for microwave foods. It discusses the types of flavor definition, the sources of natural flavours and the difference between conventional and microweave heating as well as the generation of flavour from microwave heating especially via maillard reaction. It isolates the particular effects of microwave heating on the browning reaction in flavor formation and its implications for the choice and application of flavours for microwave foods. For products that are simply seasoned and reheated, the microwave does not present significant challenges that are different from conventional heating. Microwave popcorn is an ongoing challenge to create a good flavour that will not all be volatilized during heating. Encapsulation can provide a benefit in protecting the flavour during processing and helping to retain it during the popping process. As work continues to better understand flavours, there will be new developments that will benefit microwave food products.

Author details

G.E. Ibrahim, A.H. El-Ghorab*, K.F. El-Massry and F. Osman
Chemistry of Flavour and Aroma Department, National Research Center, Tahrir St. Dokki, Cairo, Giza, Egypt

10. References

Arhne, L., Prothon, F., and F., T. (2003). Comparison of drying kinetics and texture effects of two calcium pretreatments before MW-assisted dehydration of apple and potato', Internat J Food Sci and Technol, 38(4), 411-420.

Barbiroli, G.; Garutti, A. M. and Mazzaracchio, P.((1978). Note on behavior of 1-amino-1-deoxy-2-ketose derivatives during cooking when added to starch based foodstuffs. Cereal Chem. 55:1056–1959.

Baysal, T., Ersus, S., and Icier, F. (2002). Effects of microwave and hot air combination drying on the quality of carrot', Food Sci and BioTechnol, 11(1), 19-23.

Begum, S. and Brewer, M. S. (2001). Physical, chemical and sensory quality of microwave blanched snow peas. Journal of Food Quality, 24(1), 479–493.

Boehm, M., Bade, M., and Kunz, B. (2002). Quality stabilisation of fresh herbs using a combined vacuum-microwave drying process. Adv in Food Sci, 24(2), 55-61.

* Corresponding Author

Bose, A. K.; Manhas, M. S.; Banik, B. K. and Robb, E. W. (1994). Microwave-induced organic reaction enhancement. (MORE) chemistry: Techniques for rapid, safe and inexpensive synthesis. Res. Chem. Interned. 20:1–11.

Brewer, M. S. and Begum, S. (2003). Effect of microwave power level and time on ascorbic acid content, peroxidase activity and colour of selected vegetables. Journal of Food Processing and Preservation, 27, 411–426.

Burea, M. P., Chiriife, J., Resnik, S. L. and Lozano, D. R. (1987). Nonenzymatic browning in liquid model systems of high water activity: Kinetics of colour changes due to caramelization of various sugars. J Food Sci, 52(4), 1059-1062.

Burfoot, D., Griffin, W. J. and James, S. J. (1988). Microwave pasteurization of prepared meals. J. Food Eng., 8(3), 145–156.

Burfoot, D., Railton, C. J., Foster, A. M. and Reavell, S. R. (1996). Modelling the pasteurization of prepared meals with microwaves at 896 MHz. J. Food Eng., 30(1/2), 117–133.

Calay, R.K., Newborough, M., Probert, D., and Calay, P.S. (1995). Predicative equations for the dielectric properties of foods. Int J Food Sci Technol, 29, 699-713.

Datta, A. K. and Liu, J. (1992). Thermal time distributions for microwave and conventional heating of food. Food & Bioproducts Processing, Trans. IChemE, 70, C2, 83–90.

Decareau, R. V. (1992). Microwave Foods: New Product Development, Food and Nutrition Press, Inc., Connecticut.

Decareau, R.V. (1985). Microwaves in the Food Processing Industry, Academic Press, New York.

Drake, S. R.; Spayd, S. E. and Thompson, J. B. (1981). The influence of blanch and freezing methods on the quality of selected vegetables. Journal of Food Quality, 4(4), 271–278.

Eliasson, A. C. and Larsson, K. (1993). Cereals in Breadmaking: A Molecular Colloidal Approach, Marcel Dekker, New York.

Fathima, A., Begum, K., and Rajalakshmi, D. (2001). Microwave drying of selected greens and their sensory characteristics. Plant Foods for Human Nutr, 56(4), 303-311.

Feng, H., Juming, T., and Cavalieri, R.P. (2002). Dielectric properties of dehydrated apples as affected by moisture and temperature. Transactions of the ASAE, 45(1), 129-135.

Feng, H., Juming, T., Mattinson, D.S. and Fellman, J.K. (1999). Microwave and spouted bed drying of frozen blueberries: The effect of drying and pretreatment methods on physical properties and retention of flavor volatiles. J. Food Proc and Preserv, 23(6), 463-479.

George, R. M. (1993). Making (micro)waves. Food Processing, 62(5), 23–28.

George, R. M. and Campbell, G. M. (1994). The use of metallic packaging to improve heating uniformity and process validation during microwave sterilization. In R. Field (Ed.), Food Process Engineering, IChemE Symposium, (pp. 219–225), Univ. Bath, UK.

Giese, J. (992). dvances in microwave food processing. Food Technology Special Report, 118–123.

Giguere, R. J.; Bray, T. L.; Duncan, S. M. and Majetich, G.(1986). Application of commercial microwave ovens to organic synthesis. Tetrahedron Lett. 27:4945–4948.

Guan, D., Cheng, M., Wang, Y., and Tang, J. (2004). Dielectric properties of mashed potatoes relevant to microwave and radio-frequency pasteurization and sterilization processes. J Food Sci, 69(9), 30-37.

Hallstrom, B., Skjoldebrand, C. and Tragardh, C. (1988). HeattTransfer and food products, London: Elsevier Applied Science.

Hodge, J. E. (1953). Chemistry of browning reactions in model systems, J. Agric. Food Chem., 1, 928–943.

Ibarz, A., Pagan, J. and Garza, S. (2000). Kinetic models of non-enzymatic browning in apple puree. J Sci Food Agric, 80(8), 1162-1168.

Karel, M. (1975). Physiochemical modification of the state of water in foods', in Duckworth, RB, Water Relations in Foods, Academic Press, New York.

Katz, I. (1994). Maillard, microwave and extrusion cooking: generation of aromas, in 'Thermally Generated Flavours: Maillard, Microwave and Extrusion Processes', T.H. Parliment, M.J. Morello and R. J.McGorrin (eds), American Chemical Society,Washington DC, 2-6.

Kent, M. (1987) Electric and dielectric properties of food materials. London: Science and Technology Publishers.

Keskin, S. O., Sumnu, G. and Sahin, S. (2004). Usage of enzymes in a novel baking process. Nahrung/Food, 48(2): 156-160.

Lane, R. H.; Boschung, M. D. and Abdel-Ghany, M. (1985). Ascorbic acid retention of selected vegetables blanched by microwave and conventional methods. Journal of Food Quality, 8(2&3), 139–144.

Lefort, J.F., Durance, T.D. and Upadhyaya, M.K. (2003). Effects of tuber storage and cultivar on the quality of vacuum microwave-dried potato chips. J Food Sci, 68(2), 690-696.

Lew, A., Krutzik, P. O., Hart, M. E. and Chamberlin, A. R. (2002). Increasing rates of reaction: microwave-assisted organic synthesis for combinatorial chemistry, Journal of Combinatorial Chemistry, 4(2): 95-105.

Mac-Leod, G. and Coppock, B. M. (1976). Volatile flavor components of beef boiled conventionally and by microwave radiation. J. Agric. Food Chem. 24:835–843.

Matz, S. A. (1960). Bakery Technology and Engineering, AVI Publishing, Westport, CT.

Mauron, J. (1981). The Maillard reaction in food; a critical review from the nutritional viewpoint. A, 5–35.

Moyano, P. C., Rioseco, V. K. and Gonzalez, P. A. (2002). Kinetics of crust colour changes during deep-fat frying of impregnated french fries. J Food Eng, 54: 249-255.

Mudgett, R.E. (1989). Microwave food processing: scientific status summary. Food Technol, 42(1), 117-126.

Mudgett, R.E. (1990). Developments in microwave food processing', in Schwartzberg, HG and Rao, MA, BioTechnology and Food Process Engineering, Marcel Dekker, New York, pp. 359-404.

Ohlsson, T. (1983). Fundamentals of microwave cooking, Microwave World, 4(2): 4. Code of Federal Regulations (2008). Title 21, Vol. 2, US Government Printing Office, Washington, DC.

Ozmutlu, O., Sumnu, G. and Sahin, S. (2001). Effects of different formulations on the quality of microwave baked breads. European Food Research and Technology, 213(1):38-42.

Paré, J. R.; Sigouin, M. and Lapointe, J. (1991). U.S. Pat. No 5,002,784 issued March 26.

Parliment, T. H. (1993). Comparison of thermal and microwave mediated Maillard reactions, in 'Food Flavours, Ingredients and Composition', G. Charalambous (ed), Elsevier Science Publishers, Amsterdam, 657-662.

Peterson, E. R. (1993). Microwave chemistry: A conceptual review of the literature. In Quality Enhancement Using Microwaves. 28th Annual Microwave Symposium Proceedings, International Microwave Power Institute.

Pillsbury, a division of General Mills, http://www.generalmills.com

Pomeanz, Y. and Meloan, C. E. (1987). Food Analysis: Theory and Practice, 2nd ed. Van Reinhold Nostrand, New York.

Porter,V.L. (1971). The effect of microwave finish-drying and other process factors on quality of potato chips', PhD Thesis, Univ. of Illinois, Dissertation Abs. Internat., Sect. B, Sci. and Eng., 31(9), 5414.

Prothon, F., AHRNE, L.M., Funebo, T., Kidman, S., Langton, M., and Sjoholm, I. (2002). Effects of combined osmotic and microwave dehydration of apple on texture, microstructure and rehydration characteristics. Lebensmittel-Wissenschaft und -Technol, 34(2), 95-101.

Richard, N., Gedye, F. E. and Kenneth, C. W. (1988). The rapid synthesis of organic compounds in microwave ovens. Can. J. Chem. 66:17–26.

Ryynanen, S. (1995). The electromagnetic properties of food materials. J. Food Eng., 26(4), 409–429.

Schiffman, R.F. (1986). Food product development for microwave processing. Food Technol, 40(6), 94-96.

Schiffmann, R. F. (1986). Food product development for microwave processing, Food Technology, June 1986: 94-98.

Schiffmann, R. F. (1994a). Critical factors in microwave generated aromas', in McGorrin J, Parlimant T H and Morello M J, Thermally Generated Flavors: Maillard,Microwave and Extrusion Processes, ACS Symposium Series No. 543, American Chemical Society, Washington DC, 386-394.

Schiffmann, R. F. (1994b). Critical factors in Microwave-generated aromas. In Thermally Generated Flavors: Maillard, Microwave and Extrusion Processes (J. McGorrin, T. H. Parlimant, and M. J. Morello, eds.). ACS Symposium Series, No. 543, American Chemical Society, Washington DC, pp. 386–394

Seyhun, N., Sumnu, G. and Sahin, S. (2003). Effects of different emulsifier types, fat contents, and gum types on retardation of staling of microwave-baked cakes, Nahrung/Food, 47(4): 248-251.

Sharma, G.P. and Prasad, S. (2001). Drying of garlic (Allium sativum) cloves by microwave-hot air combination. J Food Eng, 50(2), 99-105.

Shibamoto, T. and Yeo, H. (1994). Flavor in the cysteine-glucose model system prepared in microwave and conventional ovens, in Thermally Generated Flavors: Maillard, Microwave, and Extrusion Processes, T. H. Parliment, M. J. Morello, and R. J. Mc-Gorrin (eds), American Chemical Society, Washington, DC, 457–465.

Shui, T. C., Shyh, H. C. and Kung, T. W. (1990). Preparative scale organic synthesis using a kitchen microwave oven. J. Chem. Soc., Chem. Commun.11:807–809.

Steinke, J. A.; Frick, C. M. ; Gallagher, J. A. and Strassburger. K. J.(1989). Influence of Microwave heating on flavor. In Thermal Generation of Aromas (T. Parliment, R. McGorrin, and C. T. Ho, eds.). ACS Symposium Series, No. 409, American Chemical Society, Washington, DC, pp. 520–525.

Sumnu, G. (2001). A review on microwave baking of foods, International Journal of Food Science and Technology, 36: 117-127.

van Eijk, T. (1994). In Thermally Generated Flavours: Maillard, Microwave and Extrusion Processes, ed. by Parliment, T.H., Morello, M.J. and McGorrin, R.J., American Chemical Society, Washington, DC, pp. 395-404.

Venkatachalapathy, K. and Raghavan, G.S.V. (1998). Microwave drying of osmotically dehydrated blueberries', J Microwave Power and Electromag Energy, 33(2), 95-102.

Venkatachalapathy, K. and Raghavan, G.S.V. (1999). Combined osmotic and microwave drying of strawberries. Drying Technol, 17(4/5), 837-853.

Whorton, C. and Reineccius, G. (1989). Flavour development in a microwaved versus a conventionally baked cake. In Thermal Generation of Aromas (T. Parliment, R. McGorrin, and C. T. Ho, eds.). ACS Symposium Series, No. 409; American Chemical Society, Washington, DC, p. 526.

Whorton, C. and Reineccius, G. (1990). Current developments in microwave flavors', Cereal Foods World, 35(6), 553-559.

Yang, H.W. and Gunasekaran, S (2001). Temperature profiles in a cylindrical model food during pulsed microwave heating. J Food Sci, 66(7), 998-1004.

Yaylayan, V. (1997). Classification of the Maillard reaction: A conceptual approach. Trends Food Sci. Technol. 8:13–18.

Yaylayan, V. A. and Roberts, D. D. (2001). Generation and release of food aromas under microwave heating', in Datta A K and Anantheswaran R C, Handbook of Microwave Technology for Food Applications, Marcel Dekker, New York, 173-189.

Yaylayan, V.; Eorage, N. G. and Mandeville, S.(1994). Microwave and thermally induced Maillard reactions. In Thermally Generated Flavors: Maillard, Microwave and Extrusion Processes. J. McGorrin, T. H. Parlimant, and M. J. Morello (eds.). ACS Symposium Series, No. 543, American Chemical Society, Washington DC, pp. 449–456.

Yeo, H. C. H. and Shibamoto, T. J. (1991a). Microwave-induced volatiles of the Maillard model system under different pH conditions. J. Agric. Food Chem. 39:370–373.

Yeo, H. C. H. and Shibamoto, T. J. (1991b) Flavour and browning enhancement by electrolytes during microwave irradiation of the Maillard model systems. J. Agric. Food Chem. 39:948–951.

Yeo, H.C.H. and Shibamoto, T. (1991c). Chemical comparison of flavours in microwaved and conventionally heated foods. Trend FoodSci. Technol., 2, 329-332.

Yongsawatdigul, J. and Gunasekaran, S. (1996). Microwave-vacuum drying of cranberries. Part I. Energy use and efficiency', `Microwave-vacuum drying of cranberries. Part II. Quality evaluation. J Food Proc and Preserv, 20(2), 121-143.

Yousif, A.N., Scaman, C.H., Durance, T.D., and Benoit, G. (1999). Flavour volatiles and physical properties of vacuum-microwave and air-dried sweet basil (Ocimum basilicum L). J Agric Food Chem, 47, 4777-4781.

Zheng, M., Huang, Y.W., Nelson, S.O., Bartley, P.G. and Gates, K.W. (1998). Dielectric properties and thermal conductivity of marinated shrimp and channel catfish. J. Food Sci, 63(4), 668-672.

Section 2

Technology

Chapter 5

Microwave Hydrothermal and Solvothermal Processing of Materials and Compounds

Boris I. Kharisov, Oxana V. Kharissova and Ubaldo Ortiz Méndez

Additional information is available at the end of the chapter

1. Introduction

Hydrothermal synthesis (or hydrothermal method) includes the various techniques of fabrication or crystallizing substances from high-temperature aqueous solutions at high vapor pressures. In case of crystallization processes, the hydrothermal synthesis can be defined as a method of synthesis of single crystals that depends on the solubility of minerals in hot water under high pressure. The crystal growth is performed in an apparatus consisting of a steel pressure vessel called autoclave, in which a nutrient is supplied along with water. A gradient of temperature is maintained at the opposite ends of the growth chamber so that the hotter end dissolves the nutrient and the cooler end causes seeds to take additional growth.

Nowadays, combinations of different techniques are very common and the hydrothermal method is not an exception. Hydrothermal hybrid techniques are frequently applied for synthesis of materials (including nanomaterials) and chemical compounds, mainly inorganics. In order to additionally enhance the reaction kinetics or the ability to make new materials, a great amount of work has been done to hybridize the hydrothermal technique with microwaves (MW) (*microwave-hydrothermal processing)*, electrochemistry (*hydrothermal-electrochemical synthesis*), ultrasound (*hydrothermal-sonochemical synthesis)*, mechanochemistry *(mechanochemicalhydrothermal synthesis)*, optical radiation (*hydrothermal-photochemical synthesis)*, and hot-pressing (*hydrothermal hot pressing*) (Suchanek & Riman, 2006). Hydrothermal method itself, microwave-hydrothermal and microwave-solvothermal methods are, in particular, truly low-temperature methods for the preparation of nanophase materials of different sizes and shapes. These methods save energy and are environmentally friendly, because the reactions take place in closed isolated system conditions. The nanophase materials can be produced in either a batch or continuous process using the above methods. In contrast to the conventional heating hydro/solvothermal method, which requires a long time (typically half to several days) and high electric power (over a thousand Watts), microwave-assisted heating is a greener approach to synthesize materials in a shorter time (several minutes to hours) and with lower power consumption (hundreds of Watts) as a consequence of directly and uniformly heating the contents. Particular aspects of these techniques were examined in several reviews (Shangzhao Shi & Jiann-Yang Hwang, 2003; Komarneni, 2003; Komarneni & Katsuki, 2002) and a book chapter (Guiotoku et al., 2011).

In this Chapter, we try to describe briefly main aspects of hydro/solvothermal processes under simultaneous microwave heating (H-MW or S-MW). Reactions, carried out by consecutive application of hydro/solvothermal and microwave treatment, are out of scope of this study.

2. Typical equipment

Typical commercial equipment, used for MW hydro/solvothermal processing, is shown in Fig. 1. Its cost is usually about 30,000 USD. Several reports describe also home-made combinations of MW-heating and hydro/solvothermal reactions.

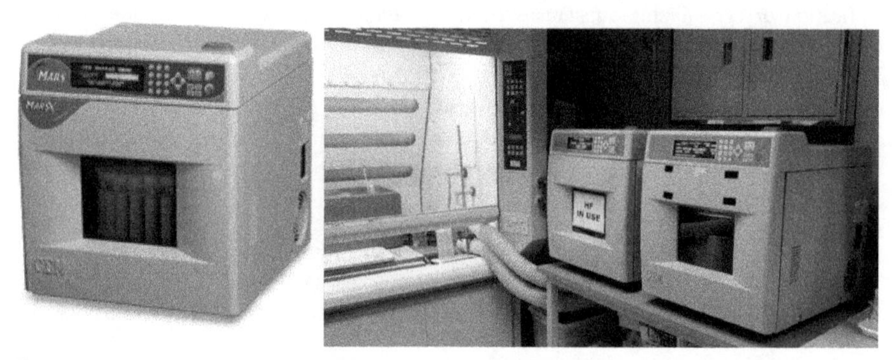

Figure 1. Typical equipment (MARS), used for MW hydro/solvothermal processing.

3. Inorganic compounds

The H-MW method was used for preparation of free or supported *elemental metals* (Cu, Ni, Co, Ag) long ago (Komarneni et al., 1995). Thus, microwave-hydrothermal processing in combination with polyol process was used to prepare Ag^{o}-, Pt^{o}- or Pd^{o}-intercalated montmorillonite (Komarneni, Hussein, et al., 1995). Sub-nanometer metal clusters were introduced into the interlayers while some larger metal particles of 5-100 nm were crystallized on the external surfaces. The small metal clusters in the interlayers and on the external surfaces may be useful in certain catalytic applications. The reduction of chlorocomplexes of gold(III) from muriatic solutions by nanocrystal powders of palladium and platinum at 110 and 130°C under H-MW conditions was studied, revealing Au-Pd and

Au-Pt bimetallic particles with a core-shell structure according to the scheme shown in Fig. 2 (Belousov et al, 2011). The obtained particles had a core of the metal reductant covered with a substitutional solid (Au, Pd) solution in case of palladium, and isolated by a gold layer in the case of platinum. It was shown by the example of the Au-Pd system that the use of microwave irradiation allowed one not only to accelerate the synthesis of particles but also to obtain more homogeneous materials in comparison with conventional heating. In addition, magnetic $FeNi_3$ nanochains were synthesized by reducing iron(III) acetylacetonate and nickel(II) acetylacetonate with hydrazine in ethylene glycol solution without any template according to the mechanism shown in Fig. 3 (Jia et al., 2010). The size of the aligned nanospheres in the magnetic $FeNi_3$ chains could be adjusted from 150 to 550 nm by increasing the amounts of the precursors. Magnetic measurement revealed that the $FeNi_3$ nanochains showed enhanced coercivity and saturation magnetization. As an example of core-shell-type gold nanoparticles, the Au/SnO_2 core-shell structure was synthesized using the H-MW method (Yu & Dutta, 2011). In MW preparation, the peak position of the UV-visible plasmon absorption band of Au nanoparticles was red shifted from 520 to 543 nm, due to the formation of an SnO_2 shell. An SnO_2 shell (thickness 10-12 nm) formation was complete within 5 min.

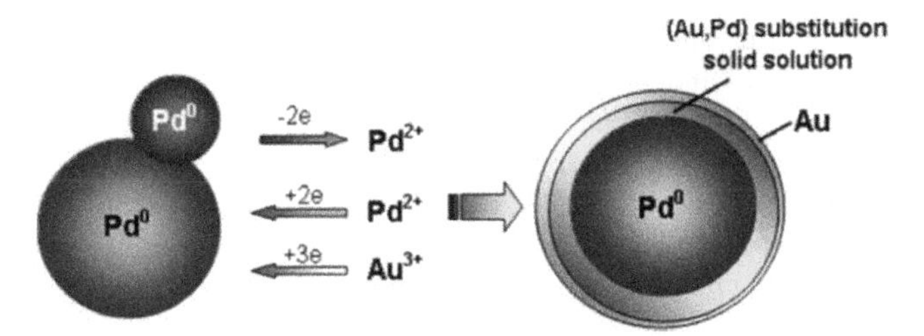

Figure 2. Core-shell Au/Pd particle formation scheme. With permission.

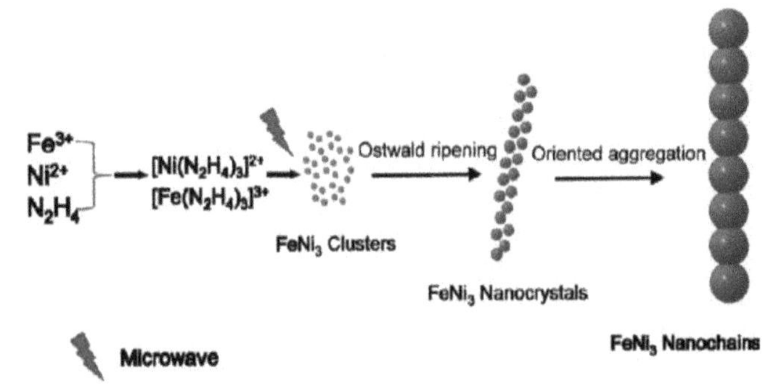

Figure 3. Illustration of a proposed mechanism for the formation of $FeNi_3$ nanochains. With permission.

Among *metal oxides*, TiO_2 is the compound, received obviously main attention of researchers due to its numerous applications, in particular in nanostructurized forms. Similar applied methods led to a variety of its distinct crystalline phases obtained in different reports. Thus, nanoparticles of *brookite*-type TiO_2 were prepared at 200°C (H-MW-heating time 5 min) for 0-60 min starting from the titanium peroxo glycolate complex in basic solution (Morishima et al., 2007). The activity in photodecomposition of oxalic acid by the samples prepared using H-MW technique was higher than activities of brookite nanoparticles prepared by the conventional hydrothermal method. On the contrary, mesoporous titanium dioxide with highly crystalline *anatase* phase and high surface area, a promising material for energy and environmental application, was obtained (Huang et al., 2011) *via H*-MW route using stable and water-soluble titanium citrate complexes as the precursors. It was shown that the synthesized TiO_2 contained mainly anatase phase with crystallite size of 5.0-8.6 nm at various hydrothermal temperatures and durations ranging from 150 to 180°C and from 30 to 120 min, respectively. The mesoporous nanocrystals synthesized at 180°C were then used to prepare the TiO_2 photoelectrode using screen-printing deposition method. The MW180-120-based TiO_2 photoanode exhibited a good efficiency on photocurrent conversion and the conversion efficiency was in the range 4.8-7.1%, depending on active area and film thickness. In addition, TiO_2 of the shuttle-like *rutile* phase (10 nm) was prepared using $TiCl_4$ and HCl by H-MW method (Chen et al., 2008). Interestingly, the use of H-MW method resulted in the formation of TiO_2 *nanotubes* comprising anatase and rutile phases (Sikhwivhilu et al., 2010). Conventional hydrothermal heating resulted in the formation of tubes with a titanate structure. The two methods yielded tubular structures with similar size dimensions, surface areas and morphologies. The two methods gave 100 % yields of tubes with different degrees of crystallinity. At last, the combination of sonication and H-MW (three methods at once) led to preparing fluorinated mesoporous TiO_2 *microspheres* (500 nm size) (Zhu et al., 2010). The authors achieved the fabrication of mesoporous TiO_2, doping of fluorine by sonication and then hydrothermal treatment of a solution containing TiO_2 precursor sol and sodium fluoride.

Zinc oxide, the principal nanotechnological object, was also intensively studied and reported in distinct forms, in particular as nanobar-structured ZnO thin film (Li et al., 2011) (obtained from Zn salt solution (nitrate, chloride, acetate, and/or sulfate) and hexamethylenetetramine solution as raw materials). The wide interest in ZnO has resulted from the following fundamental characteristic features with potential applications in electronic, structural and bio-materials: direct band gap semiconductor (3.37 eV), large excitation binding energy (60 meV), near UV emission and transparent conductivity. ZnO nanorods were synthesized using zinc nitrate and methenamine aqueous solutions in a H-MW process (Shojaee et al., 2010). It was revealed that concentration of precursors and irradiation power displayed significant influences on the compaction and dimensions of the grown nanorods. The 1D ZnO nanostructures and microstructures with a hexagonal cross-section growing in the (0002) direction were obtained under H-MW method (MW 2.45 GHz) at 130°C for 30 min (de Moura et al., 2010). In addition, the intriguing results were reported (Huang, J. et al., 2008) for a facile H-MW route employing the reaction of $Zn(NO_3)_2 \cdot 6H_2O$ and NaOH to synthesize a single-crystal zinc oxide 1D nanostructure with a 3D morphology (Fig. 4). A substantial reduction in the reaction time as well as the reaction temperature is observed compared with the hydrothermal process. Fig. 5 shows a condensed illustration of the authors' strategies in the morphology control of ZnO nanostructures. First, ZnO nuclei generally evolve into nanorods by preferential c-axis ([002] direction) oriented 1D growth. Second, nanorods can be converted into nanowires by a multiple nanorods growth along the [002] direction and simultaneous local attachment of the polar (0001) surfaces or nanospindles by an increase in diameter and local dissolution. Third, multiple nanorods grow from center results in nanodandelions. Fourth, when the crystal growth along the [002] direction is suppressed, nanoslices can be obtained due to quasi 1D growth. Finally, when multiple nanoslices grow further, nanothruster vanes can be formed by self-assembled growth. The MW-hydrothermal mechanism of ZnO nanostructures can be considered as follows (reactions 1-2):

$$Zn(NO_3)_2 + 2NaOH \rightarrow Zn(OH)_2 \downarrow + 2NaNO_3 \quad (1)$$

$$Zn(OH)_2 + 2H_2O \rightarrow Zn(OH)_4^{2-} + 2H^+ \rightarrow ZnO + 3H_2O \quad (2)$$

Figure 4. FE-SEM images of the ZnO nanocrystals with different morphologies: nanorods (a, temperature = 413 K, $[Zn^{2+}]$ = 1.6 mol L^{-1}, filling ratio = 70%, time = 20 min, middle: HRTEM image and right: selected area electron diffraction (SEAD) pattern), nanowires (b, temperature = 453 K, $[Zn^{2+}]$ = 1.6 mol L^{-1}, filling ratio = 70%, time = 20 min), nanothruster vanes (c, temperature = 393 K, $[Zn^{2+}]$ = 1.6 mol L^{-1}, filling ratio = 70%, time = 20 min), nanodandelions (d, temperature = 373 K, $[Zn^{2+}]$ = 1.6 mol L^{-1}, filling ratio = 70%, time = 20 min) and radial nanospindles (e, pressure = 3.0 MPa, $[Zn^{2+}]$ = 0.8 mol L^{-1}, filling ratio = 70%, time = 20 min). With permission.

Other simple and mixed/complex oxides are extensively reported. Thus, yttria stabilized zirconia (YSZ) is the main material for preparing functional device such as oxygen sensor, solid state oxide fuel cell and high temperature humidity transducer (Zhao et al., 2007). Its nanopowders were prepared by H-MW method with programmable MARS-5 microwave digester in strong basic media at temperature from 100-120°C and time from 1 h to 5 h, while the temperature is 190-250°C by conventional hydrothermal heating (CH). The result showed that compared with CH, H-MW can reduce the reaction time, and influence the content of product. The CH and H-MW techniques were also used for production of nanocrystalline zirconium and hafnium dioxides (8-20 nm) at 180 and 250°C and highly dispersive powders of barium zirconate and hafnate at 150°C (Maksimov et al., 2008). HfO_2 was also H-MW-obtained in a rice form (Eliziari et al., 2009) according to the following reactions 3-5:

$$HfCl_4 + 2H_2O \rightarrow Hf(OH)_2Cl_2 + 2HCl \quad (3)$$

$$Hf(OH)_2Cl_2 + 2KOH \rightarrow Hf(OH)_4 + 2KCl \quad (4)$$

$$Hf(OH)_4 \rightarrow HfO_2 + 2H_2O \quad (5)$$

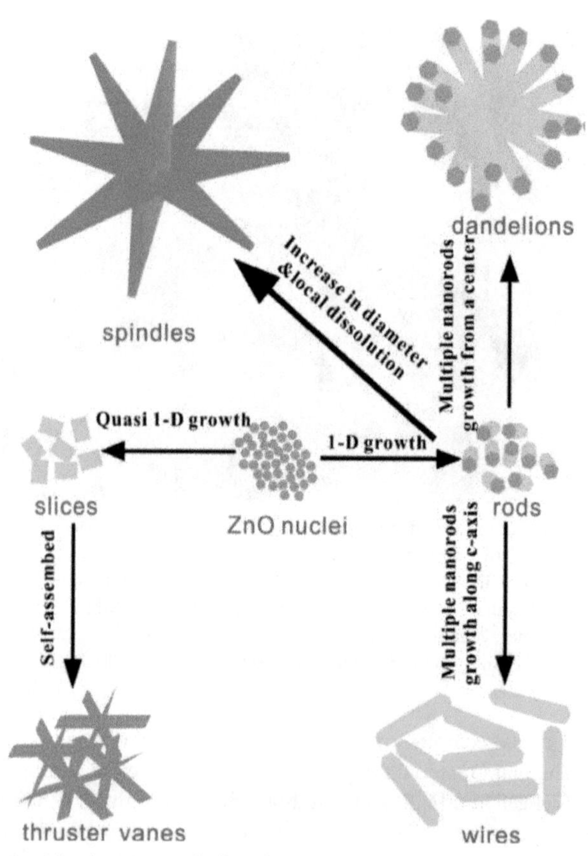

Figure 5. Schematic of the shape-controlled synthesis of ZnO nanorods, nanowires, nanothruster vanes, nanodandelions and radial nanospindles *via* a microwave hydrothermal route. With permission.

The effect of microwave radiation on the formation of smaller and uniform α-Fe_2O_3 powders from $FeCl_3$ solution at 100-140°C was investigated (Katsuki, 2009). As a practical application, a new red pigment on their basis for porcelain was developed. Structures of tin oxides in different oxidation states are known, for instance SnO_2 nanoparticles (Krishna & Komarneni, 2009) or SnO powders (Pires et al., 2008); the last ones were obtained by the H-MW technique using $SnCl_2{\cdot}2H_2O$ as a precursor. By changing the hydrothermal processing time, temperature, the type of mineralizing agent (NaOH, KOH, or NH_4OH) and its concentration, SnO crystals having different sizes and morphologies could be achieved. Plate-like form was found to be the characteristic morphology of growth. CuO with less-common sea urchin-like morphology is also known (Volanti et al., 2010). Single crystalline Co_3O_4 nanorods (Li, W.-h., 2008) or Co_3O_4 mesoporous nanowires (Zeng et al., 2011) with average single crystalline grain sizes of 8 nm, 12 nm, 25 nm, and 45 nm were synthesized by sintering the last nanostructure of H-MW-processed belt-$Co(OH)_2$ precursors at 300-500°C for 2 h. The interesting finding was made that room temperature ferromagnetism appeared at 350°C in the high orientation samples. A mixture of crystalline Co_3O_4/CoO nanorods (length of around 80 nm and an average diameter of 42 nm) with non-uniform dense distribution was synthesized by H-MW technique (Al-Tuwirqi et al., 2011). The band energy gap of the product was 1.79 eV which lies between the energy gap of CoO and that for Co_3O_4. As synthesized mixed Co_3O_4/CoO nanorods can be very useful for supercapacitor devices application. Magnetic hysteresis loops at room temperature of the as synthesized mixed oxides (Co_3O_4/CoO) nanorods exhibit typical soft magnetic behavior.

MoO_3 nanoflowers were synthesized on a Si substrate by a facility H-MW method (Wei et al., 2009). The nanoflowers consisted of tens of nanobelts and the nanobelts were about several micrometers in length, several tens to several hundreds of nanometers in width, and tens of nanometers in thickness. As-grown MoO_3 nanobelts exhibited a good field-emission property and have great potential for applications in field-emission devices. In case of tungsten oxide, its compounds with different composition were reported, for example monodisperse crystalline $WO_3{\cdot}2H_2O$ ($H_2WO_4{\cdot}H_2O$) nanospheres, which were prepared by (+)-tartaric acid-assisted H-MW process (Sun et al., 2008), meanwhile the synthesis of crystalline $W_{18}O_{49}$ with nanosheet like morphology was carried out by low cost MW irradiation method without employing hydrothermal process (Hariharan et al., 2011). The $W_{18}O_{49}$ nanosheets had the average dimensions of the order of 250 nm in length and around 150 nm in width. The band gap energies to be 3.28 and 3.47 eV for $WO_3{\cdot}H_2O$ and $W_{18}O_{49}$ samples, respectively. An hierarchically structured $WO_3{\cdot}0.33H_2O$ "snowflakes" were synthesized by a template-free and H-MW method (Li, J. et al., 2011). Their particles had an oriented growth along six equivalent <100> directions in (001) plane to form the snowflakelike microstructure, which is significantly different from the sample prepared at conventional hydrothermal conditions. Moreover, microwave heating was considered by authors to accelerate the oriented crystal growth along <100> directions. Mixed Mo-W nanostructures are also known. Thus, $W_{0.4}Mo_{0.6}O_3$ and carbon-decorated WO_x-MoO_2 (x = 2 and 3) nanorods were synthesized (Yoon & Manthiram, 2011). The carbon-decorated WO_x-MoO_2 nanorods exhibited excellent capacity retention as the carbon provides an elastic matrix for absorbing the volume expansion-contraction smoothly and prevents aggregation of the nanorods during cycling. In addition, solid solutions of titanium-tin oxide $Ti_xSn_{1-x}O_2$ were prepared by H-MW method from a solution containing $TiCl_3$ and $SnCl_4$ (Yang et al., 2011). The advantage and contribution of this technique was revealed to be effective reduction of the difference in the formation rate of TiO_2 and SnO_2, resulting in the precise control of the solid solution composition. Enlarging the crystal size of single-phase, rutile-type $Ti_xSn_{1-x}O_2$ solid solutions can be achieved by annealing in air and the crystal phase is stable at 800°C.

A series of reports is devoted to lanthanide oxides, in particular to distinct CeO_2 nanostructures, for instance virtually nonaggregated, primarily hexagonal CeO_2 nanoparticles (Ivanov et al., 2009). In another research (Dos Santos et al., 2008), crystalline CeO_2 nanoparticles were prepared by a simple and fast H-MW method at 130°C for 20 min and then were calcinated at 500°C for 1, 2 and 4 h. Ceria powders were found to have, in this case, a spherical shape with particle size below 10 nm, a narrow distribution, and exhibit weak agglomeration. In addition, ceria hollow nanospheres composed of CeO_2 nanocrystals were synthesized *via* a template-free H-MW method (Cao et al., 2010). An Ostwald ripening mechanism (Fig. 6) coupled with a self-templated, self-assembly process, in which amorphous solid spheres are converted to crystalline nanocrystals and the latter self-assemble into hollow structures, was proposed for the formation of these ceria hollow structures, which showed an excellent adsorption capacity for heavy metal ions, for example, 22.4 $mg{\cdot}g^{-1}$ for As(V) and 15.4 $mg{\cdot}g^{-1}$ for Cr(VI). The authors noted that these ceria hollow nanospheres are also excellent supports for gold nanoparticles, forming a Au/CeO_2 composite catalyst. Nanocrystalline Nd_2O_3 precursor particles were prepared by a H-MW route from a solution containing $Nd(CH_3COO)_3{\cdot}H_2O$ (Zawadzki, 2008). Further thermal treatment of the as-prepared precursors resulted in the formation of the well-crystallized Nd_2O_3 (cubic or trigonal) nanoparticles with fibrous or rod-like morphology (specific surface area 130 m^2/g).

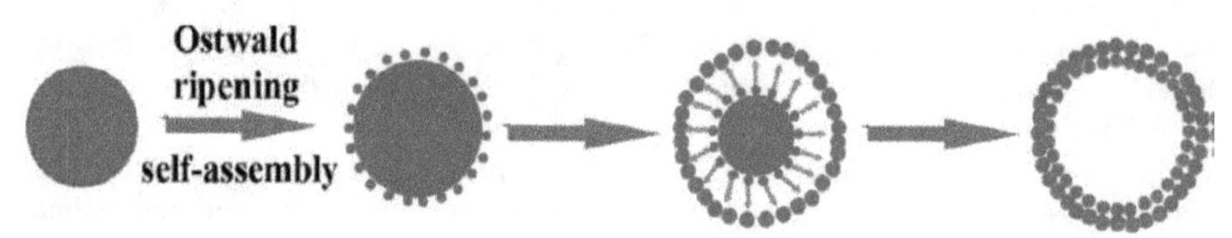

Figure 6. Illustration of the Ostwald ripening coupled self-templated, self-assembly process of the ceria precursor. With permission.

A small number of *hydroxides*, obtained by H-MW technique, are known. Thus, the CH and H-MW methods were used to synthesize layered double hydroxides (LDHs) (Wang et al., 2011). The microwave treatment LDHs (of MgAl and NiMgAl) were found to have higher crystallinity and smaller crystal sizes than the conventional hydrothermal treatment LDHs. It was indicated that the interactions of both OH^--CO_3^{2-} and CO_3^{2-}-CO_3^{2-} in NiMgAl-LDH, obtained by H-MW technique, are weaker. Also, the thermal decomposition of OH^- and CO_3^{2-} in the NiMgAl-LDH sample, obtained by H-MW technique, occurred earlier and faster than that of other LDHs. Nanostructural β-Ni hydroxide β-$Ni(OH)_2$ plates were prepared using the H-MW method at a low temperature and short reaction times (de Moura et al., 2011). An NH_3 solution was employed as the coordinating agent, which reacts with $[Ni(H_2O)_6]^{2+}$ to control the growth of β-$Ni(OH)_2$ nuclei. It was revealed that the samples consisted of hexagonal-shaped nanoplates with a different particle size distribution. Hierarchically nanostructured γ-AlOOH microspheres self-assembled by nanosheets were prepared *via* H-MW technique at 160°C for 30 min, by using $AlCl_3 \cdot 6H_2O$ and NaOH as raw materials and cetyltrimethyl ammonium bromide (CTAB) as surfactant, respectively (Liu et al., 2011). The morphology-contained γ-Al_2O_3 can be obtained through the thermal decomposition of γ-AlOOH precursors at 500°C for 2 h. Both of γ-AlOOH and γ-Al_2O_3 microspheres were used to adsorb Congo red from water solution. Cubic-shaped $In(OH)_3$ particles with average size of 0.348 μm were precipitated from a mixed aqueous solution of $InCl_3$ and urea by a H-MW method (Koga & Kimizu, 2008). No intermediate compound was found during the course of thermal decomposition from cubic-$In(OH)_3$ to cubic-In_2O_3. In addition, GaOOH nanorods were synthesized from $Ga(NO_3)_3$ *via* a facile H-MW method (Sun, M. et al., 2010). It was revealed that the as-synthesized sample was consisted of rod-like particles. The results for degradation of aromatic compounds (such as benzene and toluene) in an O_2 gas stream under UV light illumination demonstrated that GaOOH nanorods exhibited superior photocatalytic activity and stability as compared to commercial TiO_2 in both benzene and toluene degradation.

A considerable number of publications are devoted to H-MW fabrication of *oxygen-containing salts,* in particular to metal *titanates,* both simple as $BaTiO_3$ (Sun, W. et al., 2007; Nyutu et al., 2008) and more complex non-stoichiometric as $Ba_{0.75}Sr_{0.25}Zr_{0.1}Ti_{0.9}O_3$ (Chen & Luo, 2008) or $Bi_{0.5}Na_{0.5}TiO_3$. (Lv et al., 2009). Influence of reaction conditions on their synthesis and structures has been studied. Thus, the role of *in situ* stirring under H-MW conditions on the preparation of barium titanate was investigated (Komarneni & Katsuki et al., 2010). It was established that stirring under H-MW conditions in the temperature range of 150-200°C led to enhanced crystallization of Ba titanate as revealed by yields compared to the static condition. In addition, stirring led to smaller and more uniform crystals under H-MW conditions compared to those crystallized without stirring. Nanoparticles (20-40 nm) of barium titanate doped with different amounts of Sn^{2+} consisting of single phase perovskite structure were synthesized by using a S-MW reaction (Xie et al., 2009). Ceramic bodies (Fig. 7) were obtained using a spark plasma sintering method under argon atmosphere avoiding the disproportionation and oxidation of Sn^{2+} in the air. It was considered that Sn^{2+} entered into A site of the perovskite formula, ABO_3, because the lattice parameters decreased with increasing the amount of doped Sn^{2+}. The particle sizes were about 20–40 nm and increased with increasing the amount of doped Sn^{2+}. The H-MW method was used also to synthesize crystalline barium strontium titanate ($Ba_{0.8}Sr_{0.2}TiO_3$) nanoparticles (BST, 40-80 nm) in the temperature range of 100-130°C (Simoes et al., 2010). Comparing this and conventional techniques, it was shown that the H-MW synthesis route is rapid, cost effective, and could serve as an alternative to obtain BST nanoparticles. In addition, as a combination of three techniques, MW-heating was applied in the sonocatalyzed hydrothermal preparation of tetragonal phase-pure lead titanate nanopowders with stoichiometric chemical composition (Tapala et al., 2008).

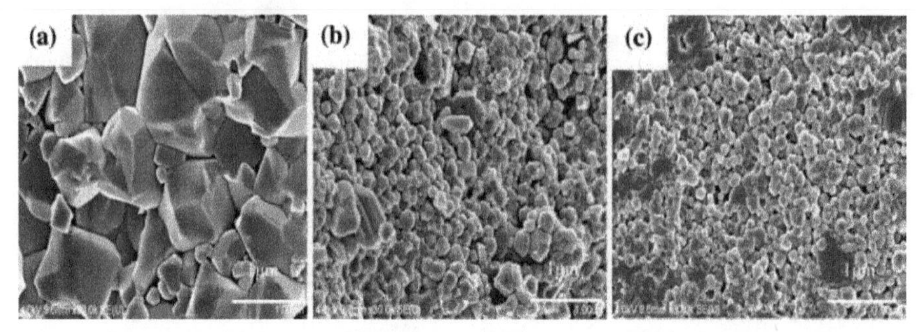

Figure 7. SEM images of a) $BaTiO_3$, b) $Ba_{0.90}Sn_{0.10}TiO_3$, and c) $Ba_{0.80}Sn_{0.20}TiO_3$ ceramic body. With permission.

Related *niobates, molybdates* (for example $SrMoO_4$ (Sczancoski et al., 2008)), *zirconates,* and *tungstates* were also widely reported. Thus, crystallization of 1D $KNbO_3$ nanostructures (generally unstable due to the natural tendency to form non-stoichiometric potassium niobates) was carried out through the reaction between Nb_2O_5 and KOH under H-MW preparation (Paula et al., 2008). The use of this method made possible a very fast preparation of single crystalline powders. Crystalline, single-phase, needle-like $Ba(Mn_{1/3}Nb_{2/3})O_3$ ceramics possessing high anisotropy were also described (Dias et al., 2009). Large crystals could be prepared from a direct combination of nanosized crystals under H-MW processing, which use leads to fast nucleation and production of nanosized particles, which could undergo a multiplying growth *via* a "cementing mechanism". It was established that the Mn ions exhibit a particular role in the lattice dynamics in complex perovskites. $BaMoO_4$ powders were prepared by the coprecipitation method and processed in a domestic H-MW equipment (Cavalcante et al., 2008). It was shown that the $BaMoO_4$ powders present a polydisperse particle size distribution, are free of secondary phases and crystallize in a tetragonal structure. In another closely related research of the same authors, octahedron-like $BaMoO_4$ microcrystals were synthesized by the same method at room temperature and processed in H-MW equipment at 413 K for different times (from 30 min to 5 h) (Cavalcante et al., 2009). The researchers revealed that as-prepared $BaMoO_4$ microcrystals present an octahedron-like morphology with agglomerate nature and poly-disperse particle size distribution. It was also indicated that the microcrystals grow along the [001] direction. A fast and economical route based on H-MW reaction was developed to synthesize pancake-like $Fe_2(MoO_4)_3$ microstructures (Zhang et al., 2010). It was established that several factors, including the amt. of nitric acid, reaction time, temperature and iron source, play crucial roles in the formation of the $Fe_2(MoO_4)_3$ multilayer stacked structures. The oriented attachment and layer-by-layer self-assembly of nanosheets is responsible for the formation of these structures. Additionally, micro-sized decaoctahedron $BaZrO_3$ powders (Fig. 8) were synthesized by means of a H-MW method at 140°C for 40 min (Moreira et al., 2009). A theoretical model derived from previous first principle calculations allowed authors to discuss the origin of the photoluminescence emission in $BaZrO_3$ powders which can be related to the local disorder in the network of both ZrO_6 octahedral and dodecahedral (BaO_{12}) hence forming the constituent polyhedron of $BaZrO_3$ system. The H-

MW processing was also employed to synthesize nanostructured alkaline-earth-metal (Ca, Ba, Sr) tungstate compounds in environmentally friendly conditions (110°C for times ranging from 5 to 20 min) (Siqueira et al., 2010).

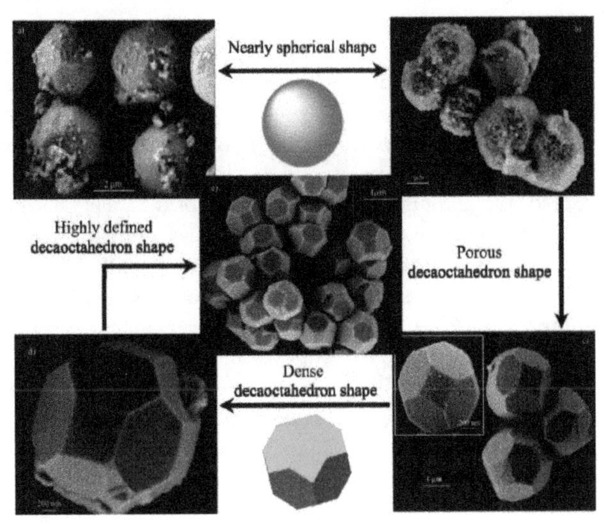

Figure 8. FE-SEM image of nearly spherical $BaZrO_3$ powders obtained for (a) 10 and (b) 20 min, and of decaoctahedral $BaZrO_3$ powders obtained for (c) 40, (d) 80, and (e) 160 min. With permission.

Ferrite salts, such as $BiFeO_3$ (Prado-Gonjal et al., 2009), and related compounds are common for H-MW synthesis. Thus, the pure phase of $BiFeO_3$ powder was synthesized by the H-MW method with $FeCl_3 \cdot 6H_2O$ and $Bi(NO_3)_3 \cdot 5H_2O$ powder as raw materials, KOH as mineralizer and Tween 80 as a surfactant (Zheng et al., 2011). It was indicated that the single phase of $BiFeO_3$ powder could be prepared *via* the H-MW method at 200°C for 1 h with KOH of 1.5 mol/L, the volume fraction of Tween 80 of 0-1.7% and a mole ratio of Fe to Bi of 1.0. When the hair ball-like $BiFeO_3$ powder was used as a catalyst, degradation rate of Rhodamine B increased from 48.92% to 79.71% after 4 h UV light irradiation. The formation of $CoFe_2O_4$ nanocrystals (particle size 6-11 nm) under hydrothermal conditions at 130°C was investigated (Kuznetsova et al., 2009). The hydrothermal medium was heated by 2 different methods, *i.e.*, the microwave (H-MW, with a reaction time from 1 min to 2 h) and conventional (with a time from 30 min to 45 h) methods. The use of microwave heating considerably accelerates the formation of $CoFe_2O_4$ particles. Preliminary ultrasonic treatment for 3 min increases the phase formation rate in the case of microwave heating and hardly affects the occurrence of the process upon conventional heating. In addition, $(Ni_{0.30}Cu_{0.20}Zn_{0.50})Fe_2O_{4\text{-}x}MnO$ (x = 0,0.01, 0.02, 0.03, 0.04) nanopowders were synthesized by H-MW method, and sintered into dense-ceramics under the conditions of 900°/4 h (Xu et al., 2007). It was established that Mn content can influence lattice parameters of samples; Mn-doping increases density of sintered Ni-Cu-Zn ferrites. Among other iron-containing compounds, we note nanocrystalline Yttrium Iron garnet (YIG) $Y_3Fe_2(FeO_4)_3$ with improved magnetic properties (Sadhana et al., 2009) and tantalum oxide-added MgCuZn ferrite powders (Krishnaveni et al., 2006).

Other oxygen-containing salts are represented by a relatively small number of examples. Thus, by using H-MW crystallization approach, $LiFePO_4$ nanoparticles were synthesized in several minutes without the use of any organic reducing agent and argon protection (Yang, G. et al., 2011). A preferential orientation of crystal growth occurs upon MW hydrothermal field. The $LiFePO_4$ crystals present 1) a change from nanoparticle to nanosheet with the increasing reaction time from 5 to 20 min, 2) a couple of redox peaks in their CV profiles, whose pairs correspond to the charge/discharge reaction of the Fe^{3+}/Fe^{2+} redox couple. The authors stated that, because of the $LiFePO_4$ samples prepared without any carbon, the initial charge/discharge capacities and cycleability of absolutely are affected by the crystal structure which is controlled by the MW irradiation condition. Flower-like $PbGeO_3$ microstructures were prepared at low temperature *via* H-MW solution-phase approach (Li, Z.-Q. et al., 2012). It was revealed that the entire structure of the architecture is composed of a large quantity of individual nanorods with 100-300 nm in width and several micrometers in length. Optical test showed that the absorption edge of the $PbGeO_3$ sample was 315 nm, corresponding to a bandgap of 3.94 eV. The $CoAl_2O_4$ pigment commonly used for coloring ceramic products was synthesized by the H-MW processing, and ink-jet printing with this aqueous pigment ink was performed to decorate porcelain (Obata et al., 2011). The synthesized $CoAl_2O_4$ particles were regular octahedrons measuring approximately 70 nm and were used to prepare an aqueous suspension, which was then used for printing on tiles by an ink-jet printing system. A convenient H-MW synthesis of nanostructured Cu^{2+}-substituted $ZnGa_2O_4$ spinels was reported (Conrad et al., 2010). A difference was observed in the coordination environments with Zn mostly situated on the tetrahedral sites of the spinel lattice whereas Cu is located on the octahedral sites of the nanostructured $ZnGa_2O_4:Cu^{2+}$ materials. Eu^{3+}, Dy^{3+} and Sm^{3+} doped nano-sized $YP_{0.8}V_{0.2}O_4$ phosphors were synthesized by a simple and facile microwave assisted hydrothermal process (Jin et al., 2011). Under UV excitation, the $YP_{0.8}V_{0.2}O_4:Ln^{3+}$ showed the VO_4^{3-} self-emission band at approximately 452 nm and the characteristic emission of doped lanthanide ions (Ln^{3+}). In addition, a facile H-MW route was developed to prepare oxynitride-based $(Sr_{1-x-y}Ce_xTb_y)Si_2O_{2-\delta}N_{2+\mu}$ phosphors (Hsu & Lu, 2011). The emitting colors of the microwave-hydrothermally derived phosphors can be tuned over a wide range under UV excitation. Hectorite (white clay mineral) $Na_{0.4}Mg_{2.7}Li_{0.3}Si_4O_{10}(OH)_2$ was prepared by aging the gel precursor by H-MW treatment at 393 K for 16 h (Vicente et al., 2009). Thus fabricated hectorite had higher purity (60%) than hectorite prepared by conventional heating (45%).

Sulfides, nitrides, and phosphides. To reduce the reaction time, electrical energy consumption, and cost, binary α-NiS-β-NiS was synthesized by a H-MW within 15 min at temperatures of 160-180°C (Idris et al., 2011). At 140°C, pure hexagonal NiAs-type α-NiS phase was identified; with increasing reaction temperature (160-180°C), an increasing fraction of rhombohedral millerite-like β-NiS is formed as a secondary phase. TEM imaging confirmed that needle-like protrusions connect the clusters of α-NiS particles.

Uniform nanorod and nanoplate structured Bi_2S_3 thin films were prepared on ITO substrates using a H-MW-assisted electrodeposition method (Wang et al., 2010). The obtained thin films are composed of orthorhombic phase Bi_2S_3 with good crystallinity. With the increase in the hydrothermal temperature (the best is 130°C), the crystallinity of the obtained films gradually improves and then decreases. In a related report (Thongtem et al., 2010), Bi_2S_3 nanorods in flower-shaped bundles were synthesized by decomposition of Bi-thiourea complexes under the microwave-assisted hydrothermal process. It was shown that Bi_2S_3 has the orthorhombic phase and appears as nanorods in flower-shaped bundles. Their UV-visible spectrum showed the absorbance at 596 nm, with its direct energy band gap of 1.82 eV. In addition, silver sulfide nanoworms (diameter of 50 nm and hundreds of nanometers in length) were prepared *via* a rapid H-MW by reacting silver nitrate and thioacetamide in the aqueous solution of the Bovine Serum Albumin (BSA) protein (Xing et al., 2011). It was shown that the nanoworms were assembled by multiple adjacent Ag_2S nanoparticles and stabilized by a layer of BSA attached to their surface. *In vitro* assays on the human cervical cancer cell line HeLa showed that the nanoworms exhibited good biocompatibility due to the presence of BSA coating. The authors stated that this combination of features makes the nanoworms attractive and promising building blocks for advanced materials and devices. As an example of practical application of metal calcogenides, obtained by H-MW technique, a red pigment, consisting of zirconium silicate-encapsulated particles of cadmium selenide sulfide, was prepared (Wang, F. et al, 2010). The GaN nanorods were synthesized by means of a combination of H-MW process and ammoniation at high temperature using Ga_2O_3 as raw material (Li, D. et al., 2011). It was found that GaN nanorods with aspect ratio of 5:1 are composed of highly oriented nanoparticles. The nanorods belong to hexagonal structure, whose crystal orientation is (002). The efficiencies of two methods of synthesizing InP micro-scale hollow spheres (Fig. 9) were compared (Xiuwen Zheng et al., 2009) *via* the analogous solution–liquid–solid (ASLS) growth mechanism, either through a traditional solvothermal procedure, or *via* a microwave-assisted method (S(H)-MW). $InCl_3{\cdot}4H_2O$, $HAuCl_4$ ethanol solution, P_4 and KBH_4 were used as precursors with ethylenediamine as solvent. MW synthesis was carried out for ca. 30 min under 180–220°C at the microwave power 600 W. For traditional solvothermal route, long time (10 h) is necessary to obtain the micrometer hollow spheres, however, for the microwave-assisted route, 30 min is enough for hollow spherical products. Fig. 10 describes the formation mechanism of InP micro-scale hollow spheres, which, according to the authors, is as follows: under the thermal MW irradiation conditions, the reaction between P_4 molecules and In is on the surface of the In/Au droplets. Thereafter, the formed InP nanoparticles undergo the solidification to form compact InP layer on the surface of the Au/In core/shell droplets, which block the further reaction of P_4 with In molecules in the beads. As a result, when removing the unreacted In by diluted HCl solution, the inner Au cores separate from the outermost InP shells, and finally produce InP hollow spheres. Due to the loss of the support, some collapsed hollow spheres are formed.

Complex fluorides (Fig. 11) with well-defined cubic morphologies KMF_3 (M = Zn, Mn, Co, Fe, materials of technological importance, were synthesized (Kramer et al., 2008; Parhi et al.,

2008) from KF and MCl_2 (M = Zn, Mn, Co, Fe) in a Parr hydrothermal vessel, subjected to MW-heating in a domestic microwave operating for 4 min at 2.45 GHz with a maximum power of 1100 W, according to the reaction (6):

$$3KF + MCl_2 \rightarrow KMF_3 + 2KCl \quad (6)$$

Figure 9. The SEM images (a and b) for **T**-InP (obtained by a **traditional** solvothermal procedure) samples and (c and d) for **M**-InP (obtained by a **microwave** solvothermal procedure) samples after HCl treatment. Inserted in (c) is the close-up of hollow nature for **M**-InP samples. With permission.

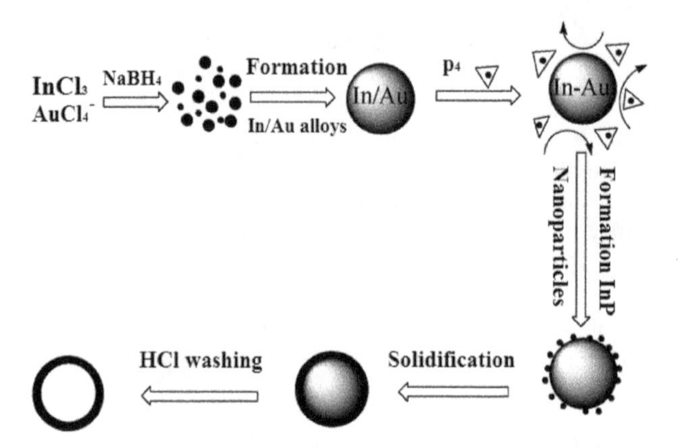

Figure 10. The proposed formation mechanism for InP hollow spheres. With permission.

Figure 11. SEM images of (a) $KZnF_3$, (b) $KMnF_3$, (c) $KCoF_3$, and (d) $KFeF_3$ synthesized by H-MW method. With permission.

4. Composites

The composites, reported as fabricated by H-MW technique, are mainly based on the simple, double oxides or salts described above. Thus, transition metal oxide/graphene composites are known (Kim et al., 2011). For single-phase unitary/binary oxides-graphene composites, a two-step strategy to prepare them is as follows: precipitation of hydroxides followed by H(S)-MW annealing (Chang et al., 2010). This method was applied to the preparation of Mn_3O_4-graphene and $NiCo_2O_4$-graphene composites. Metal oxide / CNT nano-hybrid materials ($LiMn_2O_4$/CNT, $LiCoO_2$/CNT and $Li_4Ti_5O_{12}$/CNT nanocomposites) were synthesized through selective heterogeneous nucleation and growth of the oxides on CNT surface using H-MW process (Ma et al., 2010). In the composites, CNTs acted as a substrate to deposit nano-sized metal oxide and to connect the nanoparticles along 1D conduction path. The nano-hybrid materials showed excellent high rate capability and good structural reversibility for energy storage applications. H-MW synthesis method was also applied (Guo et al., 2009) for fabrication of tin dioxide nanoparticles-multiwalled carbon nanotubes (MWCNTs) composite powder for preparation of gas sensor. The mixed solution of $SnCl_4 \cdot 5H_2O$ and NaOH at mole ratio of 1:(8-12), NaCl and multiwalled carbon nanotubes (MWCNTs) were used as reactants, subjecting to ultrasonic dispersion, and reacting at 110-180°C for 30-300 min to synthesize SnO_2 nanoparticles-MWCNTs composite powder by microwave-assisted hydrothermal reaction. Se/C nanocomposite with core-shell structures was prepared through a facile one-pot H-MW process (Yu, J.C. et al., 2005). This material, consisting of a trigonal-Se (t-Se) core and an amorphous-C (a-C) shell, can be converted to hollow carbon capsules by thermal treatment. GaP nanocrystals/morin composite fluorescent materials were prepared by microwave hydrothermal synthesis method with Na_3P, $GaCl_3$ and morin as raw materials (Cui et al., 2001). It was shown that GaP nanocrystals underwent no structure transformation under the microwave hydrothermal condition, but they grew larger after the composite reaction with morin. The wavelength of the composite materials blue-shifted and their luminescent efficiency increased when the particle size of GaP nanocrystals decreased.

Mesoporous composites of metal organic frameworks (MOFs; Cu-based) with boehmite and silica were prepared by one-pot H-MW synthesis in the presence of Pluronic-type triblock-copolymer (Gorka et al., 2010). Mesoporosity in these composites can be tailored by varying the MOF/oxide phase ratio. Fe-JLU-15 materials with different Si/Fe ratios (Si/Fe = 90, 50, 10) were synthesized by H-MW process (Bachari et al., 2009). These species correspond to hematite particles, very small "isolated" or oligomeric Fe(III) species possibly incorporated in the mesoporous silica wall, and Fe(III) oxide clusters either isolated or agglomerated, forming "rafts" at the surface of the silica and exhibiting ferromagnetic ordering. Applying similar Fe-FSM-16 materials, the liquid-phase benzylation of aromatic compounds with benzyl chloride (BC) was investigated (Bachari et al., 2010). Catalytic data in the benzylation of aromatic compounds such as benzene and toluene with BC show that Fe-FSM-16 samples synthesized by the M-H process are very active and recyclable catalysts. Nano-TiO_2 and super fine Al_2O_3 composite powder was developed by H-MW method, and TiO_2-Al_2O_3 semiconductor-dielectric composite ceramics with uniform fine grain structure were prepared by fast speed sintering in H_2 atmosphere (Lu & Zhang, 2003). Titania-hydroxyapatite (TiO_2-HAp) nanocomposite was produced by H-MW technique (Pushpakanth et al., 2008); but, in case of related Ca, Sr and $Ca_{0.5}Sr_{0.5}$ hydroxyapatites (Komarneni, Noh et al., 2010), microwave-assisted reactions did not lead to accelerated syntheses of hydroxyapatites in comparison with conventional-hydrothermal method, because the crystallization of these materials occurred at very low temperature.

In addition, the nano-sized $BaTiO_3$ and NiCuZn ferrite powders (40-60 nm) were synthesized at 160°C for 45 min (Sadhana, Praveena et al., 2009). These $xBaTiO_3$ + $(1-x)NiCuZnFe_2O_4$ nano-composites were prepared at different weight percentages. It was observed that these composites were useful for the fabrication of Multilayer Chip Inductors (MLCI). The related nanocomposites of $NiCuZnFe_2O_4$-SiO_2 (particle size of 20 nm) were prepared using H-MW method in the same conditions (Praveena et al., 2010). Among a little of organic matter-containing composites, we note the synthesis of poly (3,4-ethylenedioxythiophene)/V_2O_5 (PEDOT/V_2O_5) by *in-situ* oxidation of monomer (3,4-ethylenedioxythiophene) into crystalline nanostrip V_2O_5 using H-MW technique (Ragupathy et al., 2011). It was observed that the interlayer spacing of V_2O_5 upon intercalation of the polymer expands from 4.3 to 14°.

5. Ceramics

The hydrothermal synthesis of oxidic ceramic powders was reviewed (Somiya et al., 2005) for the decomposition of complex oxides like ilmenite, the hydrothermal oxidation of metals, hydrothermal precipitation, combinations of electrochemical, mechanical, microwave and sonochemical methods with hydrothermal methods. Niobates are widely described as ceramic basis. Thus, using KNN ("kalium" (potassium) nitrate niobate) powder prepared by the H-MW method as raw material, traditional ceramic method was employed to fabricate the KNN based lead-free piezoelectric ceramic with 1 mol.% ZnO or 1 mol.% CuO sintering additives (Li, Y. et al., 2011). The preparation method of lithium-doped potassium sodium niobate-based lead-free piezoelectric ceramic powder was based on use of MOH (M=Li, Na, K) and Nb_2O_5 as raw materials in a H-MW reactor (Tan et al., 2011). The advantages of the offered technique were low reaction temperature, short reaction period, high reaction activity of obtained powder, low energy consumption, and environmental friendliness. In addition, $K_{0.5}Na_{0.5}NbO_3$ lead-free piezoelectric ceramics were prepared from their powders which were synthesized by H-MW method (Zhou et al., 2010). The results indicated that phase of KNN ceramics were pure orthorhombic symmetry. When the powders synthesized at 160°C for 7 h, the final ceramic grains possess considerably better distribution and more homogonous in size.

Among other ceramics, indium vanadates for severe applications as photocatalysts, anodes for Li rechargeable batteries or electrochromic devices were prepared *via* H-MW synthesis performed at 220°C for different reaction times (Bartonickova et al., 2010). The H-MW method was applied to the preparation of strontium-doped lanthanum manganites with different stoichiometric ratio of the three oxides, $La_{1-x}Sr_xMnO_3$ (x = 0.3, 0.5, 0.6) (Rizzuti & Leonelli, 2009). The complete chemistry, mineralogical and microstructural characterization of the powders revealed the same structural properties of the perovskite powders previously synthesized by ceramic and conventional hydrothermal routes. In addition, a series of $Ni_{0.5-x}Cu_xZn_{0.5}Fe_2O_4$ (x = 0.05, 0.10, 0.15, and 0.20) ferrites nanopowders were synthesized by H-MW method, and sintered into dense-ceramics under the conditions of 900°C/4 h (Zheng, Ya-lin et al., 2007). The performed studies showed within a limited Cu content range of x = 0.05-0.20, copper ions were present in different ionic states in the A- and B-sites which could influence the size of lattice parameter. It was also found that crystallite size, initial permeability, resistivity and quality factor were the highest when the Cu content was 0.20.

6. Metal complexes

There are obviously no many examples of coordination compounds, obtained by H(S)-MW technique due to low stability of organic matter in hydrothermal conditions. Thus, an organic-inorganic UV absorber $[Hgua]_2 \cdot (Ti_5O_5F_{12})$ was obtained (Lhoste et al., 2011), whose 3D network is built up from infinite inorganic layers $\infty(Ti_5O_5F_{12})$ separated by guanidinium cations. Under UV irradiation at 254 nm for 40 h, the white microcrystalline powder turned to light purple-gray due to reduction of Ti(IV) to Ti(III), confirmed by magnetic measurements. Two extended solids displaying both 1D coordination polymer $[Co(H_2O)_4(4,4'\text{-bipy})](4,4'\text{-bipyH}_2)_2(SO_4) \cdot 2H_2O$ (bipy = 2,2'-bipyridine) and 2D H-bonded structural features $[Co_2(4,4'\text{-bipy})_2(SO_4)_2(H_2O)_6] \cdot 4(H_2O)$ were prepared (Prior et al., 2011) under H-MW conditions. Within the first framework is located a twice protonated 4,4'-bipyridine molecule ($C_{10}N_2H_{10}^{2+}$) which forms 2 short N-H······O H-bonds and 8 further non-classical C-H·····O interactions. The second compound displays 1D chains of Co-bipyridine which are sinusoidal in nature. Three mixed-ligand Co(II) complexes (Shi et al., 2009)

$[Na_2Co(\mu_4\text{-btec})(H_2O)_8]_n$, $[Co_2(\mu_2\text{-btec})(bipy)_2(H_2O)_6]{\cdot}2H_2O$, and $[Co_2(\mu_2\text{-btec})(phen)_2(H_2O)_6]{\cdot}2H_2O$ (H_4btec = 1,2,4,5-benzenetetracarboxylic acid, phen = 1,10-phenanthroline) are also known. A vanadium 2,6-naphthalenedicarboxylate, $V^{III}(OH)(O_2C\text{-}C_{10}H_6\text{-}CO_2){\cdot}H_2O$ was synthesized under S-MW procedure (Liu, Y.-Ya et al., 2012). After calcination at 250°C in air, the V^{III} center was oxidized to V^{IV} with the structure of $V^{IV}O(O_2C\text{-}C_{10}H_6\text{-}CO_2)$. The last compound, in the liquid-phase oxidation of cyclohexene, exhibited catalytic performance similar to $[VO(O_2C\text{-}C_6H_4\text{-}CO_2)]$. The compound is reusable and maintains its catalytic activity through several runs. Purinium, adeninium, and guaninium fluoroaluminates, $[Hpur]_2{\cdot}(AlF_5)$, $[Hade]_3{\cdot}(AlF_6){\cdot}6.5H_2O$ and $[Hguan]_3{\cdot}(Al_3F_{12})$, were synthesized by H-MW synthesis at 120°C or 190°C (Cadiau et al., 2011). Authors commented that the crystallization was difficult; all crystals of the first two complexes were very small while only a microcrystalline powder of the third compound was obtained. The purine, adenine, and guanine amines were found to be monoprotonated and lie between the preceding chains or layers.

By treating $Cu(NO_3)_2{\cdot}3H_2O$ with a V-shaped ligand 4,4′-oxydibenzoic acid (H_2oba), a dynamic metal-carboxylate framework $[Cu_2(oba)_2(DMF)_2]{\cdot}5.25DMF$ (MCF-23, Fig. 12) was synthesized, which features a wavelike layer with rhombic grids based on the paddle-wheel secondary building units (Xiao-Feng Wang et al., 2008). MCF-23 synthesized by conventional solvothermal methods always contained considerable and intractable impurities. In contrast, a S-MW method was proven to be a faster and greener approach to synthesize phase-pure MCF-23 in high yield. Larger crystals suitable for single-crystal diffraction could be obtained by the multistep microwave heating mode. Also, a 3D coordination copper polymer, $[Cu_2(pyz)_2(SO_4)(H_2O)_2]_n$ (pyz = pyrazine), was synthesized under H-MW conditions (Amo-Ochoa et al., 2007). The authors especially note that microwave assisted synthesis produces monocrystal suitable for X-ray diffraction studies, reducing reaction time and with higher yield than the classical hydrothermal procedures.

Figure 12. SEM images of MCF-23 synthesized *via* single-step microwave-assisted solvothermal synthesis (MASS): (a) 1 min, (b) 5 min, (c) 10 min, (d) 30 min, (e) 150 min at 160 °C, and (f) photo of single crystals synthesized *via* multistep MASS. With permission.

Single crystals of [H_3dien]·(FeF_6)·H_2O and [H_3dien]·(CrF_6)·H_2O were obtained by S-MW (Ben Ali et al., 2007). These structures are built up from isolated FeF_6 or CrF_6 octahedra, water molecules and triprotonated amines; each octahedron is connected by hydrogen bonds to six organic cations and two water molecules. ^{57}Fe Mossbauer spectrometry proved the hyperfine structure confirming the presence of Fe^{3+} in octahedral coordination and reveals the existence of paramagnetic spin fluctuations. Under H-MW conditions, two compounds Zn(pinH)(H_2O) and Cd(pinH) were obtained through the reactions of Zn (or Cd) sulfate and 2-phosphonic-isonicotinic acid (pinH_3) (Yang Yi-F. et al., 2007). The authors established that the first complex has a layer structure in which double chains composed of corner-sharing {ZnO_5N} octahedra and {CPO_3} tetrahedra are cross-linked through the carboxylate groups. The latter has a framework structure where the inorganic layers composed of edge-sharing {CdO_5N} octahedra and {CPO_3} tetrahedra are pillared by the pyridyl carboxylate groups. A 3D open-framework and a chain-structured zinc terephthalate were obtained by hydrothermal crystallization under microwave heating at 180°C (Rajic et al., 2006). These structures have rather similar principal building units, which facilitates the structure transformation from the low-dimensional to the open-framework one during crystallization. For the optimization of the H-MW synthesis of $[Zn(BDC)(H_2O)_2]_n$, where H_2BDC = 1,4-benzenedicarboxylic acid, the reactions were carried out at the fixed temperature of 120°C for 10, 20, 30 and 40 min (Wanderley et al., 2011). Pure crystalline $[Zn(BDC)(H_2O)_2]_n$ was obtained in high yield (90%) with a reaction time of 10 min.

7. Other materials and processes

Series of reports are devoted to H-MW fabrication of a variety of *molecular sieves* and other *adsorbents*. Thus, a range of nanosize alkaline-free gallosilicate mesoporous molecular sieves (GaMMS) were synthesized using H-MW method (Cheng et al., 2011). These nanosize GaMMSs exhibited high surface area (240-720 m^2/g), pore volume (1.06-1.49 m^3/g), narrow pore size distribution and nano-particle size between 20 and 100 nm and four-coordinated gallium site mainly. It was revealed that the nanosize GaMMS shows a much higher activity than that of conventional GaMCM-41, probably due to higher concentrations of H-form sites, external surface and fast diffusion. Mesoporous molecular sieves MCM-41 modified by single (Ti) and bimetal (Ti-V) ions with highly ordered hexagonal arrangement of their cylindrical channels were prepared by direct synthesis under H-MW conditions at 403 K (Guo, Y. et al., 2010). It was shown that Ti and V ions were introduced into MCM-41 under M-H conditions and Ti/V-Si bond was formed. The modified materials were high active and selective in the epoxidation of styrene at 343 K in comparison with single functional MCM-41. The applied method greatly improved the selectivity to styrene oxide. In addition, a rapid process to prepare cryptomelane-type octahedral molecular sieve (OMS-2) nanomaterials using a H-MW technique (MW-HT) for 10 s (in comparison with up to 4 days in a conventional hydrothermal reaction) was presented (Huang, H. et al., 2010). It was shown that the OMS-2 nanowires were produced from thin nano-flakes with increasing reaction temperatures. Carbon/silica adsorbents (carbosils) were prepared by pyrolisis of CH_2Cl_2 at 823 K and the reaction time from 0.5 to 6 h on the mesoporous silica gel surface and then hydrothermally treated at 473 K with steam or liquid water by using the classical autoclave with traditional heating way or in the microwave reactor (Skubiszewska-Zieba, 2008). It was stated that hydrothermal treatment in the microwave reactor, contrary to that in the classical autoclave, allowed obtaining adsorbents with noticeably higher values of total pore volume in relation to the initial adsorbents and in majority with a higher specific surface area. Application of microwave energy allowed obtaining adsorbents with lower values of surface free energy in relation to the initial adsorbents and those modified in the autoclave.

Carbon-based materials are also produced by H-MW method. Thus, rice chaff can be converted into activated carbon by alkaline treatment at high temperature, in particular by H-MW treatment with NaOH solution (Inada et al., 2011). Rice chaff was heated at 700°C for 1 h in N_2 after HCl treatment to remove impurity, such as alkali metals, resulting in the formation of charcoal. After further necessary steps, it was revealed that the specific surface area of product was 1060 m^2/g under microwave irradiation and 690 m^2/g by conventional heating with autoclave, which indicated that H-MW treatment is effective for activation of charcoal. In addition, acidic suspensions of multi-walled carbon nanotubes (MWCNTs) in 5 M HNO_3 solutions were readily obtained by microwave-assisted hydrothermal digestion and their rapid surface functionalization with polyaniline could be achieved growing composite coatings *in situ* onto graphite electrodes in an acidified suspension of MWCNTs and aniline (Wu et al., 2007). The advantage of the integrity of the MWCNTs after microwave exposure was used to improve the mechanical strength of the composite coatings, especially when an utmost thick coating was attempted for a high capacitance.

Materials and processes using iron-containing compounds. Iron-FSM [folded-sheet mesoporous material]-16 materials with different Si/Fe ratio (Si/Fe = 90, 60, and 10) were synthesized by H-MW process (Bachari et al., 2011). These species correspond to hematite particles, small isolated Fe(III) species possibly incorporated in the mesoporous silica wall, and iron oxide clusters either isolated or agglomerated, forming rafts at the surface of the silica and exhibiting ferromagnetic ordering. It was shown that Fe-FSM-16 synthesized this way are active and recycle catalysts. The high yield and rapid precipitation of Fe(III) arsenate(V) dihydrate (identical to the mineral scorodite) from aqueous Fe(III)-As(V) solutions at pH 0.6-1.1 was achieved long ago using microwave dielectric heating (Baghurst et all., 1995). The process could be useful for As removal from industrial wastes. Metallic iron (α-Fe)/manganese-zinc ferrite ($Fe_{3-x-y}Mn_xZn_yO_4$) nanocomposites were synthesized for 15 s using H(S)-MW treatment of alcohol solutions of chloride precursors and sodium ethoxide (Caillot et al., 2011). For all samples, 20% of metallic iron was routinely obtained using the microwave flash synthesis. Consequently, the microwave heating appears to provide an efficient source of energy in producing metallic iron nanoparticles protected against oxidation by an oxide matrix. Additionally, as an improvement of the conventional hydrothermal reaction with iron powder, NaOH and H_2O as reactants at 423 K, leading to iron oxide Fe_3O_4, $NaFeO_2$ and hydrogen, MW-heating was adopted to induce the hydrothermal reaction (Liu, X. et al., 2011). Under MWs, NaOH and H_2O absorbed microwave energy by space charge polarization and dipolar polarization and instantly converted it into thermal energy, which initiated the hydrothermal reaction that involved with zero-valent iron. The developed microwave-hydrothermal reaction was employed for the dechlorination of PCBs (polychlorinated biphenyls). For PCBs in 10 mL simulative transformer oil, almost complete dechlorination was achieved by 750 W MW irradiation for 10 min, with 0.3 g iron powder, 0.3 g NaOH and 0.6 mL H_2O added. The authors stated that MW irradiation combined with the common and cheap materials, iron powder, NaOH and H_2O, might provide a fast and cost-effective method for the treatment of PCBs-containing wastes.

Zeolites/alumosilicates, composites and *membranes* on their basis are frequently (especially ZSM-5) obtained by the discussed method. Thus, evaluation of hydrothermal synthesis of the zeolite BZSM-5 was performed by treating the synthesis mixture by different aging processes, namely, ultrasonic, static, stirring, and microwave-assisted aging prior to the conventional hydrothermal treatment (Abrishamkar et al., 2010). That the ultrasonic and microwave assisted aging shortened the crystallization time and altered the crystal size and the morphology of the obtained products. Mullite ($Al_6Si_2O_{13}$) powders were prepared by sol-gel combined with H-MW process using tetra-Et orthosilicate and $Al(NO_3)_3 \cdot 9H_2O$ as raw materials (Yang, Q. et al., 2010). It was shown that the increasing of microwave hydrothermal temperature can decrease the synthetic temperature of mullite. The initial temperature of the mullitization of gels was about 1046°C, but the temperature of preparation of pure mullite should be at 1300°C, which can be decreased to 1200°C through H-MW process. Using silatrane and alumatrane templates, remarkably uniform zeolites (NaA, EDI-type, ABW-type, FAU-type, K, and ZSM-5 zeolites) were obtained via the sol-gel process and H-MW treatment to obtain small and uniform zeolite crystals to prepare zeolite membranes (Wongkasemjit, 2009). The organic ligands of both precursors not only make the molecules stable to moisture, but also, after hydrolysis, provide trialkanolamine molecules which could function as another template for the reaction. The formation of ZSM-5 zeolite/porous carbon composite from carbonized rice husk was investigated using microwave- and conventional-hydrothermal reaction at 140-160°C (Katsuki et al., 2005). It was established that, compared to the conventional-hydrothermal (C-H) formation of ZSM-5 zeolite, the H-MW reaction led to increased rate of formation by 3-4 times at 150°C. The surface area of ZSM-5 zeolite (without template)/porous carbon composite was shown to be 485.4 m^2/g and this composite had both micropores and mesopores. Zeolite nanocrystals (e.g., silicalite-1, ZSM-5, LTL, BEA and LTA) with controllable size, morphology and SiO_2/Al_2O_3 ratio were prepared (Yuanyuan Hu et al., 2009). It is found that high synthesis temperature and long reaction time benefit the growth of all the referred zeolite nanocrystals. The authors established that for the nanozeolites crystallized in low alkalinity systems (e.g., ZSM-5, BEA and LTA), both increasing the alkalinity and decreasing the water content accelerate their nucleation process and thereby result in the decrease of their crystal size. On the contrary, for those prepared in high alkalinity systems (e.g., LTL and silicalite-1), an inversed trend could be observed. In addition, stable zeolite beta coatings with a thickness of 1-2 μm were synthesized on a borosilicate glass substrate by H-MW synthesis (Muraza et al., 2008). In addition, a regular nanocrystalline supramolecular Mg-Al hydrotalcite was prepared *via* glycol-frequency H-MW reaction using $MgCl_2 \cdot 6H_2O$,

$AlCl_3 \cdot 6H_2O$ as raw material and Na_2CO_3 as precipitator (Wu, J. et al., 2010). It was shown that the hydrotalcite exhibited as a homogeneous and hexagonal sheet.

A high quality pure hydroxy-sodalite $Na_8[AlSiO_4]_6(OH)_2$ zeolite membrane was H-MW-synthesized on an a-Al_2O_3 support (Xiaochun Xu et al., 2004). This process only needed 45 min and synthesis was more than 8 times faster than by the conventional hydrothermal synthesis method. The pure hydroxy-sodalite zeolite membrane method was found to be well inter-grown and the thickness of the membrane was 6–7 μm. Gas permeation results showed that the hydrogen/n-butane permselectivity of the hydroxy-sodalite zeolite membrane was larger than 1000, being a promising candidate for the separation of hydrogen from gas mixtures and important for the emerging hydrogen energy fuel system. The potential of microwave heating for the rapid synthesis of thin silicalite-1 membranes by secondary growth from microwave-derived silicalite-1 seeds was evaluated (Motuzas et al. 2006). The morphology, thickness, homogeneity, crystal preferential orientation and single gas permeation properties of the silicalite-1 membranes were studied in relation to the synthesis parameters. Other related zeolite-ceramic membranes (titanosilicalite TS-1 (Sebastian, V.; Motuzas, J. et al., 2010) and MFI-type (Sebastian, V.; Mallada, R. et al., 2010)) are also known.

Other H-MW-synthesized materials and compounds are rare; some of them also found useful practical applications, as, for example, titanium-based nanometer pigments (Paskocimas et al., 2009). CdS/Titanate nanotubes (TNTs) were also prepared (Chen, Y.-C. et al., 2011). It was established that the CdS nanoparticles synthesized using a 140-W microwave irradiation power at 423 K photodegraded 26% ammonia in water, while the photocatalytic efficiency increased to 52.3% using the synthesized CdS/TNTs composites. So, it can be stated that the CdS/TNTs photocatalysts possess improved photocatalytic activity than that of CdS or TNTs materials alone. Mesoporous composites of metal organic frameworks (MOFs; Cu-based) with boehmite and silica were prepared by one-pot H-MW synthesis in the presence of Pluronic-type triblock-copolymer (Gorka, J.; Fulvio, P.F. et al., 2010). A variety of visible-light-driven silver vanadates, including α-$AgVO_3$, β-$AgVO_3$, and α-Ag_3VO_4, were synthesized using a H-MW synthesis method (Pan et al., 2011). The α-Ag_3VO_4 crystalline sample with rich hydroxyl functional groups on the surface exhibited the highest degree of photocatalytic activity. Thus, the reaction rates of the photodegradation of isopropanol (IPA) and benzene vapors were approximately 8 times higher than those of P25 under visible-light irradiation. In addition, the photocatalytic activities of H-MW samples were higher than those of samples produced by conventional hydrothermal techniques. The authors explained this due to an increase in the specific surface area and additional hydroxyl functional groups on the surface. Another application of the H-MW method for destruction of organic pollutants was reported in (Liu, X. et al., 2011). The reaction of reduced iron powder and NaOH or KOH led to obtaining iron oxides and/or ferrites, releasing hydrogen for quick dechlorination of persistent organic pollutants. The method was shown to have the advantages of short treatment time and high dechlorination efficiency and can be used for treating waste organochlorine pesticides, polychlorobiphenyl oil in transformer, soil and sediment heavily polluted by organochlorine pesticides and polychlorobiphenyl oil, and garbage burning fly ash and chlor-alkali brine sludge containing high-concentration dioxins. $Ca(OH)_2$ alone and $Ca(OH)_2$ with H_3PO_4 addition (P-addition) were effectively used to remove and recover boron from wastewater using hydrothermal methods (Tsai et al., 2011). A microwave hydrothermal method was also used and compared with the conventional heating method in batch experiments. For the case of $Ca(OH)_2$ alone and the MW method, expterimental results showed that boron recovery efficiency reached 90% within 10 min, and crystals of $Ca_2B_2O_5 \cdot H_2O$ were observed. For the case of P-addition and the MW method, boron recovery efficiency reached 99% within 10 min, and Ca phosphate species ($CaHPO_4 \cdot H_2O$, $CaHPO_4$ and $Ca_{10}(PO_4)_6(OH)_2$) were formed. Hydrophobic organoclays (hybrids derived from an ion exchange of hydrophilic clays with quaternary ammonium salts and used as rheological additives in paints, inks, cosmetics, nanocomposites, and as pollutant absorbing agents in soil remediation programs) were studied using a natural (Na-montmorillonite) and several synthetic clays (Na-fluorophlogopites) as precursors (Baldassari). These organoclays synthesized using both conventional hydrothermal and H-MW processes.

8. Conclusions

Nowadays, the H(S)-MW techniques are already classic method for fabrication of distinct chemical compounds and materials, including nanomaterials. Inorganic compounds and materials are generally obtained by this route, although a certain number of organic-containing compounds are also reported. Use of organic matter in this method can be considered as a careful pioneer experimentation in order to check its suitability for MW-hydro(solvo)thermal reactions. Compounds, which are stable in hydrothermal conditions, such as metal oxides, oxygen-containing and other metal salts, a variety of zeolites, carbon-based materials, as well as composites on their basis, are classic synthesis objects by this route. It is expected that more organic/organometallic products could be tried to be prepared in H(S)-MW conditions, for example thermally stable (up to 500-600°C, that is rare for organic matter) aromatic macrocycles of phthalocyanine (Edrissi et al., 2007) type and related compounds.

Author details

Boris I. Kharisov, Oxana V. Kharissova and Ubaldo Ortiz Méndez
Universidad Autónoma de Nuevo León, Monterrey, México

9. References

Abrishamkar, M.; Azizi, S.N.; Kazemian, H. (2010). Ultrasonic-Assistance and Aging Time Effects on the Zeolitation Process of BZSM-5 Zeolite. *Zeitschrift fuer Anorganische und Allgemeine Chemie*, 636(15), 2686-2690.

Al-Tuwirqi, R.M.; Al-Ghamdi, A.A.; Al-Hazmi, F.; Alnowaiser, F.; Al-Ghamdi, A.A.; Aal, N.A.; El-Tantawy, F. (2011). Synthesis and physical properties of mixed Co_3O_4/CoO nanorods by microwave hydrothermal technique. *Superlattices and Microstructures*, 50(5), 437-448.

Amo-Ochoa, P.; Givaja, G.; Miguel, P.J.S; Castillo, O.; Zamora, F. (2007). Microwave assisted hydrothermal synthesis of a novel CuI-sulfate-pyrazine MOF. *Inorganic Chemistry Communications*, 10(8), 921-924.

Bachari, K.; Lamouchi, M. (2009). Synthesis and Characterization of Fe-JLU-15 Mesoporous Silica Via a Microwave-Hydrothermal Process. *Journal of Cluster Science*, 20(3), 573-586.

Bachari, K.; Guerroudj, R. M.; Lamouchi, M. (2011). High activities of iron-FSM-16 materials synthesized by a microwave-hydrothermal process in Friedel-Crafts alkylations. *Kinetics and Catalysis*, 52(1), 119-127.

Bachari, K.; Guerroudj, R. M.; Lamouchi, M. (2010). Catalytic performance of iron-mesoporous nanomaterials synthesized by a microwave-hydrothermal process. *Reaction Kinetics, Mechanisms and Catalysis*, 100(1), 205-215.

Baghurst, D.R.; Barrett, J.; Mingos, D.M.P. (1995). The hydrothermal microwave synthesis of scorodite: iron(III) arsenate(V) dihydrate, $FeAsO_4{\cdot}2H_2O$. *Journal of the Chemical Society, Chemical Communications*, (3), 323-324.

Baldassari, S.; Komarneni, S.; Mariani, E.; Villa, C. (2006). Microwave versus conventional preparation of organoclays from natural and synthetic clays. *Applied Clay Science*, 31(1-2), 134-141.

Bartonickova, E.; Cihlar, J. (2010). Synthesis and processing of $InVO_4$ ceramics. *International Journal of Modern Physics B: Condensed Matter Physics, Statistical Physics, Applied Physics*, 24(6&7), 770-779.

Belousov, O.V.; Belousova, N.V.; Sirotina, A.V.; Solovyov, L.A.; Zhyzhaev, A.M.; Zharkov, S.M.; Mikhlin, Y.L. (2011). Formation of Bimetallic Au-Pd and Au-Pt Nanoparticles under Hydrothermal Conditions and Microwave Irradiation. *Langmuir*, 27(18), 11697-11703.

Ben Ali, A.; Dang, M.T.; Greneche, J.-M.; Hemon-Ribaud, A.; Leblanc, M.; Maisonneuve, V. (2007). Synthesis, structure of $[H_3dien]{\cdot}(MF_6){\cdot}H_2O$ (M = Cr, Fe) and ^{57}Fe Moessbauer study of $[H_3dien]{\cdot}(FeF_6){\cdot}H_2O$. *Journal of Solid State Chemistry*, 180(6), 1911-1917.

Cadiau, A.; Adil, K.; Hemon-Ribaud, A.; Leblanc, M.; Jouanneaux, A.; Slawin, A. M. Z.; Lightfoot, P.; Maisonneuve, V. (2011). Fluoroaluminates of purine and DNA bases, adenine, guanine. $[Hpur]_2{\cdot}(AlF_5)$, $[Hade]_3{\cdot}(AlF_6){\cdot}6.5H_2O$, $[Hguan]_3{\cdot}(Al_3F_{12})$. *Solid State Sciences*, 13(1), 151-157.

Caillot, T.; Pourroy, G.; Stuerga, D. (2011). Novel metallic iron/manganese-zinc ferrite nanocomposites prepared by microwave hydrothermal flash synthesis. *Journal of Alloys and Compounds*, 509(8), 3493-3496.

Cao, C.-Y.; Cui, Z.-M.; Chen, C.-Q.; Song, W.-G.; Cai, W. (2010). Ceria Hollow Nanospheres Produced by a Template-Free Microwave-Assisted Hydrothermal Method for Heavy Metal Ion Removal and Catalysis. *Journal of Physical Chemistry C*, 114(21), 9865-9870.

Cavalcante, L. S.; Sczancoski, J. C.; Tranquilin, R. L.; Joya, M. R.; Pizani, P. S.; Varela, J. A.; Longo, E. (2008). $BaMoO_4$ powders processed in domestic microwave-hydrothermal:

Synthesis, characterization and photoluminescence at room temperature. *Journal of Physics and Chemistry of Solids*, 69(11), 2674-2680.

Cavalcante, L.S.; Sczancoski, J.C.; Tranquilin, R.L.; Varela, J.A.; Longo, E.; Orlandi, M.O. (2009). Growth mechanism of octahedron-like $BaMoO_4$ microcrystals processed in microwave-hydrothermal: experimental observations and computational modeling. *Particuology*, 7(5), 53-362.

Chang, K.-H.; Lee, Y.-F.; Hu, C.-C.; Chang, C.-I.; Liu, C.-L.; Yang, Yi-L. (2010). A unique strategy for preparing single-phase unitary/binary oxides-graphene composites. *Chemical Communications*, 46(42), 7957-7959.

Chen, J.; Luo, K. (2008). Synthesis of $Ba_{0.75}Sr_{0.25}Zr_{0.1}Ti_{0.9}O_3$ nano-powder by microwave-hydrothermal method. *Taoci* (Xianyang, China), (10), 28-30.

Chen, Y.-C.; Lo, S.-L.; Ou, H.-H.; Chen, C.-H. (2011). Photocatalytic oxidation of ammonia by cadmium sulfide/titanate nanotubes synthesized by microwave hydrothermal method. *Water Science and Technology*, 63(3), 550-557.

Chen, Z.; Li, W.; Zeng, W.; Li, M.; Xiang, J.; Zhou, Z.; Huang, J. (2008). Microwave hydrothermal synthesis of nanocrystalline rutile. *Materials Letters*, 62(28), 4343-4344.

Cheng, C.-F.; Liu, S.-M.; Cheng, H.-H.; Yao, M.G.; Liu, S.B. (2011). Microwave hydrothermal synthesis and acidity of nanosize gallosilicate mesoporous molecular sieves. *Journal of the Chinese Chemical Society*, 58(2), 155-162.

Conrad, F.; Zhou, Y.; Yulikov, M.; Hametner, K.; Weyeneth, S.; Jeschke, G.; Guenther, D.; Grunwaldt, J.-D.; Patzke, G.R. (2010). Microwave-Hydrothermal Synthesis of Nanostructured Zinc-Copper Gallates. *European Journal of Inorganic Chemistry*, (13), 2036-2043.

Cui, D.; Wei, J.; Pan, J.; Hao, X.; Xu, X.; Jiang, M. (2001). Preparation and characterization of GaP nano-composite fluorescent material. *Gongneng Cailiao*, 32(5), 543-545.

Dias, A.; Matinaga, F.M.; Moreira, R.L. (2009). Vibrational Spectroscopy and Electron-Phonon Interactions in Microwave-Hydrothermal Synthesized $Ba(Mn_{1/3}Nb_{2/3})O_3$ Complex Perovskites. *Journal of Physical Chemistry B*, 113(29), 9749-9755.

Dos Santos, M. L.; Lima, R. C.; Riccardi, C. S.; Tranquilin, R. L.; Bueno, P. R.; Varela, J. A.; Longo, E. (2008). Preparation and characterization of ceria nanospheres by microwave-hydrothermal method. *Materials Letters*, 62(30), 4509-4511.

Edrissi, M.; Nasernejad, B.; Sayedi, B. (2007). Novel method for the preparation of copper phthalocyanine blue nanoparticles in an electrochemical cell irradiated by microwave. *IJE Transactions B: Applications*, 20(3), 257-262.

Eliziari, S.A.; Cavalcante, L.S.; Sczancoski, J.C.; Pizani, A.P.C.; Varela, J.A.; Espinosa, J.W.M.; Longo, E. (2009). Morphology and Photoluminescence of HfO_2 Obtained by Microwave-Hydrothermal. *Nanoscale Res. Lett.*, 4, 1371-1379.

Gorka, J.; Fulvio, P.F.; Pikus, S.; Jaroniec, M. (2010). Mesoporous metal organic framework-boehmite and silica composites. *Chemical Communications*, 46(36), 6798-6800.

Gorka, J.; Fulvio, P.F.; Pikus, S.; Jaroniec, M. (2010). Mesoporous metal organic framework-boehmite and silica composites. *Chemical Communications*, 46(36), 6798-6800.

Guiotoku, M.; Maia, C. M. B. F.; Rambo, C.R.; Hotza, D. (2011). Chapter 13. Synthesis of Carbon-Based Materials by Microwave Hydrothermal Processing. in *"Microwave Heating"*, Edited By: Usha Chandra, INTECH, pp. 291-308.

Guo, M.; Wang, Y.; Zhang, M.; Wang, X.; Yue, C. (2009). Microwave-assisted hydrothermal synthesis method for tin dioxide nanoparticles-multiwalled carbon nanotubes (MWCNTs) composite powder for preparation of gas sensor. 7 pp., CN 101439855.

Guo, Y.; Wang, G.; Wang, Y.; Liu, Z.; Liu, G.; Liu, Y. (2010). Microwave hydrothermal synthesis of bimetallic (Ti-V) ions modified MCM-41 for epoxidation of styrene. *Materials Research Society Symposium Proceedings*, 1279(New Catalytic Materials), Paper #26.

Hariharan, V.; Parthibavarman, M.; Sekar, C. (2011). Synthesis of tungsten oxide ($W_{18}O_{49}$) nanosheets utilizing EDTA salt by microwave irradiation method. *Journal of Alloys and Compounds*, 509(14), 4788-4792.

Hsu, C.-H.; Lu, C.-H. (2011). Microwave-hydrothermally synthesized ($Sr_{1-x-y}Ce_xTb_y$)$Si_2O_{2-\delta}N_{2+\delta}$ phosphors: efficient energy transfer, structural refinement and photoluminescence properties. *Journal of Materials Chemistry*, 21(9), 2932-2939.

Huang, C.-H.; Yang, Y.-T.; Doong, R.-A. (2011). Microwave-assisted hydrothermal synthesis of mesoporous anatase TiO_2 via sol-gel process for dye-sensitized solar cells. *Microporous and Mesoporous Materials*, 142(2-3), 473-480.

Idris, N.H.; Rahman, Md.M.; Chou, S.-L.; Wang, J.-Z.; Wexler, D.; Liu, H.-K. (2011). Rapid synthesis of binary α-NiS-β-NiS by microwave autoclave for rechargeable lithium batteries. *Electrochimica Acta*, 58, 456-462.

Inada, M.; Koga, T.; Tanaka, Y.; Enomoto, N.; Hojo, J. (2011). Synthesis of activated carbon from rice chaff by microwave hydrothermal method. *Funtai oyobi Funmatsu Yakin*, 58(10), 598-601.

Ivanov, V.K.; Polezhaeva, O.S.; Gil', D.O.; Kopitsa, G.P.; Tret'yakov, Yu.D. (2009). Hydrothermal Microwave Synthesis of Nanocrystalline Cerium Dioxide. *Doklady Chemistry*, 426(2), 131–133.

Jia, J.; Yu, J.C.; Wang, Yi-X.J.; Chan, K.M. (2010). Magnetic Nanochains of $FeNi_3$ Prepared by a Template-Free Microwave-Hydrothermal Method. *ACS Applied Materials & Interfaces*, 2(9), 2579-2584.

Jin, Yu; Li, C.; Xu, Z.; Cheng, Z.; Wang, W.; Li, G.; Lin, J. (2011). Microwave-assisted hydrothermal synthesis and multicolor tuning luminescence of $YP_xV_{1-x}O_4$:Ln^{3+} (Ln = Eu, Dy, Sm) nanoparticles. *Materials Chemistry and Physics*, 129(1-2), 418-423.

Huang, H.; Sithambaram, S.; Suib, S. (2010). Microwave-assisted hydrothermal synthesis of cryptomelane-type octahedral molecular sieves (OMS-2) and their catalytic studies. Abstracts of Papers, *240th ACS National Meeting*, Boston, MA, United States, August 22-26, 2010, CATL-48.

Huang, J.; Xia, C.; Cao, L.; Zeng, X. (2008). Facile microwave hydrothermal synthesis of zinc oxide one-dimensional nanostructure with three-dimensional morphology. *Materials Science and Engineering B*, 150, 187-193.

Katsuki, H.; Furuta, S.; Watari, T.; Komarneni, S. (2005). ZSM-5 zeolite/porous carbon composite: Conventional- and microwave-hydrothermal synthesis from carbonized rice husk. *Microporous and Mesoporous Materials*, 86(1-3), 145-151.

Katsuki, H. (2009). Sintering of α-Fe_2O_3 particles prepared from microwave-hydrothermal reaction. *Funtai oyobi Funmatsu Yakin*, 56(12), 738-743.

Kim, G.B.; Kim, H.G.; Kim, Ji.Y.; Park, S.H. (2011). Preparation method of transition metal oxide/graphene composite using microwave-hydrothermal process and its use in energy-storage devices and gas sensors. 17 pp., KR 2011121584.

Koga, N.; Kimizu, T. (2008). Thermal decomposition of indium(III) hydroxide prepared by the microwave-assisted hydrothermal method. *Journal of the American Ceramic Society*, 91(12), 4052-4058.

Komarneni, S.R; Pidugu, Quing Hua Li; Roy, R. (1995). Microwave-hydrothermal processing of metal powders. *J. Mater. Res.*, 10(7), 1687-1692.

Komarneni, S.; Hussein, M. Z.; Liu, C.; Breval, E.; Malla, P. B. (1995). Microwave-hydrothermal processing of metal clusters supported in and/or on montmorillonite. *European Journal of Solid State and Inorganic Chemistry*, 32(7/8), 837-849.

Komarneni, S. (2003). Nanophase materials by hydrothermal, microwave-hydrothermal, and microwave-solvothermal methods. *Current Science*, 85(12), 1730-1734.

Komarneni, S.; Katsuki, H. (2002). Nanophase materials by a novel microwavehydrothermal process. *Pure Appl. Chem.*, 74(9), 1537–1543.

Komarneni, S.; Katsuki, H. (2010). Microwave-hydrothermal synthesis of barium titanate under stirring condition. *Ceramics International*, 36(3), 1165-1169.

Komarneni, S.; Noh, Y.D.; Kim, J.Y.; Kim, S.H.; Katsuki, H. (2010). Solvothermal/hydrothermal synthesis of metal oxides and metal powders with and without microwaves. *Zeitschrift fuer Naturforschung, B: A Journal of Chemical Sciences*, 65(8), 1033-1037.

Kramer, J.W.; Parhi, P.; Manivannan, V. (2008). Microwave Initiated Hydrothermal Synthesis of Nano-Sized Complex Fluorides, KMF_3 (Zn, Mn, Co and Fe). Abstracts, 43rd Midwest Regional Meeting of the American Chemical Society, Kearney, NE, United States, October 8-11, 2008, MWRM-319.

Krishna, M.; Komarneni, S. (2009). Conventional- vs microwave-hydrothermal synthesis of tin oxide, SnO_2 nanoparticles. *Ceramics International*, 35(8), 3375-3379.

Krishnaveni, T.; Murthy, S. R.; Gao, F.; Lu, Q.; Komarneni, S. (2006). Microwave hydrothermal synthesis of nanosize Ta_2O_5 added Mg-Cu-Zn ferrites. *Journal of Materials Science*, 41(5), 1471-1474.

Kuznetsova, V. A.; Almjasheva, O. V.; Gusarov, V. V. (2009). Influence of microwave and ultrasonic treatment on the formation of $CoFe_2O_4$ under hydrothermal conditions. *Glass Physics and Chemistry*, 35(2), 205-209.

Lhoste, J.; Rocquefelte, X.; Adil, K.; Dessapt, R.; Jobic, S.; Leblanc, M.; Maisonneuve, V.; Bujoli-Doeuff, M. (2011). A New Organic-Inorganic Hybrid Oxyfluorotitanate $[Hgua]_2\cdot(Ti_5O_5F_{12})$ as a Transparent UV Filter. *Inorganic Chemistry*, 50(12), 5671-5678.

Li, C.; Yu, R.; Ren, T.; Zhang, W. (2011). Microwave hydrothermal-deposition synthesis method for directional nanobar-structured ZnO thin film. 13 pp. CN 102260046.

Li, D.; Wang, F.; Zhu, J.-f.; Xue, X.-s. (2011). Synthesis and characterization of GaN nanorods by microwave hydrothermal method and treatment with ammonia. *Gongneng Cailiao Yu Qijian Xuebao*, 17(2), 218-222.

Li, J.; Huang, J.; Yu, C.; Wu, J.; Cao, L.; Yanagisawa, K. (2011). Hierarchically structured snowflakelike $WO_3 \cdot 0.33H_2O$ particles prepared by a facile, green, and microwave-assisted method. *Chemistry Letters*, 40(6), 579-581.

Li, W.-h. (2008). Microwave-assisted hydrothermal synthesis and optical property of Co_3O_4 nanorods. *Materials Letters*, 62(25), 4149-4151.

Li, Z.-Q.; Zhang, L.; Lin, X.-S.; Chen, X.-T.; Xue, Zi-L. (2012). Fast preparation of flower-like $PbGeO_3$ microstructures at low-temperature via a microwave-assisted hydrothermal process. *Materials Letters*, 68, 344-346.

Li, Y.; Liu, H.; Shen, Z.; Hong, Y.; Wang, Z.; Li, R. (2011). Effect of ZnO and CuO sintering additives on the properties of KNN piezoelectric ceramic. *Zhongguo Taoci*, 47(10), 28-31.

Liu, X.; Zhao, W.; Zhang, G.; Sun, K.; Zhao, Y. (2011). Method for treating persistent organic pollutants through microwave hydrothermal reaction. 6 pp., CN 102068782.

Liu, Y.-Ya; Leus, K.; Grzywa, M.; Weinberger, D.; Strubbe, K.; Vrielinck, H.; Van Deun, R.; Volkmer, D.; Van Speybroeck, V.; Van Der Voort, P. (2012). Synthesis, Structural Characterization, and Catalytic Performance of a Vanadium-Based Metal-Organic Framework (COMOC-3). *European Journal of Inorganic Chemistry*, Ahead of Print http://onlinelibrary.wiley.com/doi/10.1002/ejic.201101099/abstract

Liu, H.; He, X.-m.; Li, G.-j.; Cai, Zi-j.; Zhu, Z.-f. (2011). Microwave hydrothermal synthesis of AlOOH and Al_2O_3 hierarchically nanostructured microspheres self-assembled by nanosheets. *Gongneng Cailiao*, 42(5), 854-857, 861.

Liu, X.; Zhao, W.; Sun, K.; Zhang, G.; Zhao, Y. (2011). Dechlorination of PCBs in the simulative transformer oil by microwave-hydrothermal reaction with zero-valent iron involved. *Chemosphere*, 82(5), 773-777.

Lu, Y.; Zhang, J. (2003). Development of TiO_2-Al_2O_3 semiconductor-dielectric composite ceramics and investigation on its microwave-absorbing properties. *Xiyou Jinshu Cailiao Yu Gongcheng*, 32(Suppl. 1), 463-466.

Lv, Y.; Liu, Y.; Dai, S.; Shi, S. (2009). Synthesis of $Bi_{0.5}Na_{0.5}TiO_3$ spherical particles by microwave hydrothermal process. 7 pp., CN 101525239.

Ma, S.B.; Kim, J.G.; Kim, H.-K.; Choi, H.-R.; Kim, K.-B. (2010). Metal oxide/carbon nanotubes nano-hybrid materials for energy storage applications. Abstracts of Papers, *240th ACS National Meeting*, Boston, MA, United States, August 22-26, 2010, FUEL-52.

Maksimov, V. D.; Meskin, P. E.; Churagulov, B. R. (2008). Hydrothermal microwave synthesis of finely divided powders of simple and complex zirconium and hafnium oxides. *Poverkhnost*, (2), 76-82.

Moreira, M.L.; Andres, J.; Varela, J.A.; Longo, E. (2009). Synthesis of Fine Micro-sized $BaZrO_3$ Powders Based on a Decaoctahedron Shape by the Microwave-Assisted Hydrothermal Method. *Crystal Growth & Design*, 9(2), 833-839.

Morishima, Y.; Kobayashi, M.; Petrykin, V.; Kakihana, M.; Tomita, K. (2007). Microwave-assisted hydrothermal synthesis of brookite nanoparticles from a water-soluble titanium complex and their photocatalytic activity. *Journal of the Ceramic Society of Japan,* 115(Dec.), 826-830.

Motuzas, J.; Julbe, A.; Noble, R. D.; van der Lee, A.; Beresnevicius, Z. (2006). Rapid synthesis of oriented silicalite-1 membranes by microwave-assisted hydrothermal treatment. *J. Microporous and Mesoporous Materials,* 92(1-3), 259-269.

de Moura, A. P.; Lima, R. C.; Moreira, M. L.; Volanti, D. P.; Espinosa, J. W. M.; Orlandi, M. O.; Pizani, P. S.; Varela, J. A.; Longo, E. (2010). ZnO architectures synthesized by a microwave-assisted hydrothermal method and their photoluminescence properties. *Solid State Ionics,* 181(15-16), 775-780.

de Moura, A. P.; Lima, R. C.; Paris, E. C.; Li, M. S.; Varela, J. A.; Longo, E. (2011). Formation of nickel hydroxide plate-like structures under mild conditions and their optical properties. *Journal of Solid State Chemistry,* 184(10), 2818-2823.

Muraza, O.; Rebrov, E.V.; Chen, J.; Putkonen, M.; Niinistoe, L.; de Croon, M.H.J. M.; Schouten, J.C. (2008). Microwave-assisted hydrothermal synthesis of zeolite Beta coatings on ALD-modified borosilicate glass for application in microstructured reactors. *Chemical Engineering Journal,* 135(Suppl. 1), S117-S120.

Nyutu, E.K.; Chun-Hu Che; Dutta, P.K.; Suib, S.L. (2008). Effect of Microwave Frequency on Hydrothermal Synthesis of Nanocrystalline Tetragonal Barium Titanate. *J. Phys. Chem. C,* 112, 9659–9667.

Obata, S.; Kato, M.; Yokoyama, H.; Iwata, Y.; Kikumoto, M.; Sakurada, O. (2011). Synthesis of nano $CoAl_2O_4$ pigment for ink-jet printing to decorate porcelain. *Journal of the Ceramic Society of Japan,* 119(Mar.), 208-213.

Pan, G.-T.; Lai, M.-H.; Juang, R.-C.; Chung, T.-W.; Yang, T.C.-K. (2011). Preparation of Visible-Light-Driven Silver Vanadates by a Microwave-Assisted Hydrothermal Method for the Photodegradation of Volatile Organic Vapors. *Industrial & Engineering Chemistry Research,* 50(5), 2807-2814.

Parhi, P.; Kramer, J.; Manivannan, V. (2008). Microwave initiated hydrothermal synthesis of nano-sized complex fluorides, KMF_3 (M = Zn, Mn, Co, and Fe). *Journal of Materials Science,* 43(16), 5540-5545.

Paskocimas, C.A.; Longo da Silva, E.; Volanti, D.P.; Varela, J.A.; Silva Junior, W.; Silva, W.; Boeing, E. (2009). Process for preparation of titanium-based nanometer pigments by hydrothermic preparation aided with microwaves. Braz. Pedido PI, 19 pp., BR 2008001228.

Paula, A.J.; Parra, R.; Zaghete, M.A.; Varela, J.A. (2008). Synthesis of $KNbO_3$ nanostructures by a microwave assisted hydrothermal method. *Materials Letters,* 62(17-18), 2581-2584.

Pires, F. I.; Joanni, E.; Savu, R.; Zaghete, M. A.; Longo, E.; Varela, J. A. (2008). Microwave-assisted hydrothermal synthesis of nanocrystalline SnO powders. *Materials Letters,* 62(2), 239-242.

Prado-Gonjal, J.; Villafuerte-Castrejon, M. E.; Fuentes, L.; Moran, E. (2009). Microwave-hydrothermal synthesis of the multiferroic $BiFeO_3$. *Materials Research Bulletin*, 44(8), 1734-1737.

Praveena, K.; Sadhana, K.; Murthy, S.R. (2010). Microwave-hydrothermal synthesis of $Ni_{0.53}Cu_{0.12}Zn_{0.35}Fe_2O_4/SiO_2$ nanocomposites for MLCI. *Integrated Ferroelectrics*, 119, 122-134.

Prior, T.J.; Yotnoi, B.; Rujiwatra, A. (2011). Microwave synthesis and crystal structures of two cobalt-4,4'-bipyridine-sulfate frameworks constructed from 1-D coordination polymers linked by hydrogen bonding. *Polyhedron*, 30(2), 259-268.

Pushpakanth, S.; Srinivasan, B.; Sreedhar, B.; Sastry, T.P. (2008). An *in situ* approach to prepare nanorods of titania-hydroxyapatite (TiO_2-HAp) nanocomposite by microwave hydrothermal technique. *Materials Chemistry and Physics*, 107(2-3), 492-498.

Ragupathy, P.; Vasan, H.N.; Munichandraiah, N.; Vasanthacharya, N. (2011). *In-situ* preparation of PEDOT/V_2O_5 nanocomposite and its synergism for enhanced capacitive behavior. *Proceedings of SPIE*, 8035 (Energy Harvesting and Storage: Materials, Devices, and Applications II), 80350I/1-80350I/11.

Rajic, N.; Stojakovic, D.; Logar, N. Zabukovec; Kaucic, V. (2006). An evidence for a chain to network transformation during the microwave hydrothermal crystallization of an open-framework zinc terephthalate. *Journal of Porous Materials*, 13(2), 153-156.

Rizzuti, A.; Leonelli, C. (2009). Microwave advantages in inorganic synthesis of $La_{0.5}Sr_{0.5}MnO_3$ powders for perovskite ceramics. *Processing and Application of Ceramics*, 3(1-2), 29-32.

Sadhana, K.; Shinde, R.S.; Murthy, S.R. (2009). Synthesis of nanocrystalline YIG using microwave-hydrothermal method. *International Journal of Modern Physics B: Condensed Matter Physics, Statistical Physics, Applied Physics*, 23(17), 3637-3642.

Sadhana, K.; Praveena, K.; Bharadwaj, S.; Murthy, S.R. (2009). Microwave-Hydrothermal synthesis of $BaTiO_3$+$NiCuZnFe_2O_4$ nanocomposites. *Journal of Alloys and Compounds*, 472(1-2), 484-488.

Sczancoski, J.C.; Cavalcante, L.S.; Joya, M.R.; Varela, J.A.; Pizani, P.S.; Longo, E. (2008). $SrMoO_4$ powders processed in microwave-hydrothermal: Synthesis, characterization and optical properties. *Chemical Engineering Journal*, 140, 632–637.

Sebastian, V.; Mallada, R.; Coronas, J.; Julbe, A.; Terpstra, R.A.; Dirrix, R.W.J. (2010). Microwave-assisted hydrothermal rapid synthesis of capillary MFI-type zeolite-ceramic membranes for pervaporation application. *Journal of Membrane Science*, 355(1-2), 28-35.

Sebastian, V.; Motuzas, J.; Dirrix, R.W.J.; Terpstra, R.A.; Mallada, R.; Julbe, A. (2010). Synthesis of capillary titanosilicalite TS-1 ceramic membranes by MW-assisted hydrothermal heating for pervaporation application. *Separation and Purification Technology*, 75(3), 249-256.

Shangzhao Shi; Jiann-Yang Hwang. (2003). Microwave-assisted wet chemical synthesis: advantages, significance, and steps to industrialization. *Journal of Minerals & Materials Characterization & Engineering*, 2(2), 101-110.

Shi, Z.-F.; Jin, J.; Li, L.; Xing, Y.-H.; Niu, S.-Y. (2009). Syntheses, structures, and surface photoelectric properties of Co-btec complexes. *Wuli Huaxue Xuebao*, 25(10), 2011-2019.

Shojaee, N.; Ebadzadeh, T.; Aghaei, A. (2010). Effect of concentration and heating conditions on microwave-assisted hydrothermal synthesis of ZnO nanorods. *Materials Characterization*, 61(12), 1418-1423.

Sikhwivhilu, L.M.; Mpelane, S.; Moloto, N.; Ray, S.S. (2010). Hydrothermal synthesis of TiO_2 nanotubes: microwave heating versus conventional heating. *Ceramic Engineering and Science Proceedings*, 31(7, Nanostructured Materials and Nanotechnology IV), 45-49.

Simoes, A. Z.; Moura, F.; Onofre, T. B.; Ramirez, M. A.; Varela, J. A.; Longo, E. (2010). Microwave-hydrothermal synthesis of barium strontium titanate nanoparticles. *Journal of Alloys and Compounds*, 508(2), 620-624.

Siqueira, K.P.F.; Moreira, R.L.; Valadares, M.; Dias, A. (2010). Microwave-hydrothermal preparation of alkaline-earth-metal tungstates. *Journal of Materials Science*, 45(22), 6083-6093.

Skubiszewska-Zieba, J. (2008). Structural and energetic properties of carbosils hydrothermally treated in the classical autoclave or the microwave reactor. *Adsorption*, 14(4/5), 695-709.

Somiya, S.; Roy, R.; Komarneni, S. (2005). Hydrothermal synthesis of ceramic oxide powders. *Materials Engineering*, 28(Chemical Processing of Ceramics (2nd Edition)), 3-20.

Suchanek, W.L.; Riman, R.E. (2006). Hydrothermal Synthesis of Advanced Ceramic Powders. *Advances in Science and Technology*, *45, 184-193.*

Sun, M.; Li, D.; Zhang, W.; Fu, X.; Shao, Y.; Li, W.; Xiao, G.; He, Y. (2010). Rapid microwave hydrothermal synthesis of GaOOH nanorods with photocatalytic activity toward aromatic compounds. *Nanotechnology*, 21(35), 355601/1-355601/7.

Sun, Q.; Luo, J.; Xie, Z.; Wang, J.; Su, X. (2008). Synthesis of monodisperse $WO_3 \cdot 2H_2O$ nanospheres by microwave hydrothermal process with (+)-tartaric acid as a protective agent. *Materials Letters*, 62(17-18), 2992-2994.

Sun, W.; Pang, Y.; Li, J.; Ao, W. (2007). Particle Coarsening II: Growth Kinetics of Hydrothermal $BaTiO_3$. Chemistry of Materials, 19(7), 1772-1779.

Tan, G.; Xiong, P.; Qin, B. (2011). Method for preparing lithium-doped potassium sodium niobate-based lead-free piezoelectric ceramic powder by microwave hydrothermal method. 7 pp., CN 102205988.

Tapala, S.; Thammajak, N.; Laorattanakul, P.; Rujiwatra, A. (2008). Effects of microwave heating on sonocatalyzed hydrothermal preparation of lead titanate nanopowders. *Materials Letters*, 62(21-22), 3685-3687.

Thongtem, T.; Pilapong, C.; Kavinchan, J.; Phuruangrat, A.; Thongtem, S. (2010). Microwave-assisted hydrothermal synthesis of Bi_2S_3 nanorods in flower-shaped bundles. *Journal of Alloys and Compounds*, 500(2), 195-199.

Tsai, H.-C.; Lo, S.-L. (2011). Boron removal and recovery from concentrated wastewater using a microwave hydrothermal method. *Journal of Hazardous Materials*, 186(2-3), 1431-1437.

Vicente, I.; Salagre, P.; Cesteros, Y.; Guirado, F.; Medina, F.; Sueiras, J.E. (2009). Fast microwave synthesis of hectorite. *Applied Clay Science*, 43(1), 103-107.

Volanti, D.P.; Orlandi, M.O.; Andres, J.; Longo, E. (2010). Efficient microwave-assisted hydrothermal synthesis of CuO sea urchin-like architectures via a mesoscale self-assembly. *Cryst. Eng. Comm.*, 12(6), 1696-1699.

Wanderley, K.A.; Alves, S., Jr.; Paiva-Santos, C. de Oliveira. (2011). Microwave-assisted hydrothermal synthesis as an efficient method for obtaining $[Zn(BDC)(H_2O)_2]_n$ metal-organic framework. *Quimica Nova*, 34(3), 434-438.

Wang, F.; Liu, D.; Li, Q.; Li, D.; Zhu, J. (2010). Method for preparation of red pigment consisting of zirconium silicate-encapsulated particles of cadmium selenide sulfide by microwave hydrothermal process. 9 pp., CN 101786902.

Wang, L.; Li, B.; Yang, M.; Chen, C.; Liu, Y. (2011). Effect of Ni cations and microwave hydrothermal treatment on the related properties of layered double hydroxide-ethylene vinyl acetate copolymer composites. *Journal of Colloid and Interface Science*, 356(2), 519-525.

Wang, Y.; Huang, J.-F.; Zhu, H.; Cao, Li-Y.; Xue, X.-S.; Zeng, X.-R. (2010). Preparation of Bi_2S_3 thin films by microwave-hydrothermal assisted electrodeposition method. *Wuji Huaxue Xuebao*, 26(6), 977-981.

Wei, G.; Qin, W.; Zhang, D.; Wang, G.; Kim, R.; Zheng, K.; Wang, L. (2009). Synthesis and field emission of MoO_3 nanoflowers by a microwave hydrothermal route. *Journal of Alloys and Compounds*, 481(1-2), 417-421.

Wongkasemjit, S. (2009). Novel route to remarkably uniform zeolites. Editor(s): Wongkasemjit, S.; Jamieson, A.M. *Advanced Metal and Metal Oxide Technology*, 1-18. Publisher: Transworld Research Network, Trivandrum, India.

Wu, J.; Liang, H.; Xiao, Y.; Lin, J. (2010). Glycol-frequency microwave-hydrothermal synthesis and characterization of excellent quality Mg-Al hydrotalcite. *Zhongshan Daxue Xuebao, Ziran Kexueban*, 49(3), 70-74.

Wu, M.; Zhang, L.; Wang, D.; Gao, J.; Zhang, S. (2007). Electrochemical capacitance of MWCNT/polyaniline composite coatings grown in acidic MWCNT suspensions by microwave-assisted hydrothermal digestion. *Nanotechnology*, 18(38), 385603/1-385603/7.

Xiao-Feng Wang; Yue-Biao Zhang; Hong Huang; Jie-Peng Zhang; Xiao-Ming Chen. (2008). Microwave-Assisted Solvothermal Synthesis of a Dynamic Porous Metal-Carboxylate Framework. *Cryst. Growth & Design*, 8(12), 4559–4563.

Xiaochun Xu; Yun Bao; Chunshan Song; Weishen Yang; Jie Liu; Liwu Lin. (2004). Microwave-assisted hydrothermal synthesis of hydroxy-sodalite zeolite membrane. *Microporous and Mesoporous Materials*, 75, 173-181.

Xie, Y.; Yin, S.; Hashimoto, T.; Kimura, H.; Sato, T. (2009). Microwave-hydrothermal synthesis of nano-sized Sn^{2+}-doped $BaTiO_3$ powders and dielectric properties of corresponding ceramics obtained by spark plasma sintering method. *Journal of Materials Science*, 44(18), 4834-4839.

Xing, R.; Liu, S.; Tian, S. (2011). Microwave-assisted hydrothermal synthesis of biocompatible silver sulfide nanoworms. *Journal of Nanoparticle Research*, 13(10), 4847-4854.

Xiuwen Zheng; Qitu Hua; Chuansheng Sun. (2009). Efficient rapid microwave-assisted route to synthesize InP micrometer. *Mater. Res. Sci.*, 44(1), 216-219.

Xu, G.-l.; Zheng, Ya-l.; Lai, Z.-yu. (2007). Effects of Mn on the properties of Ni-Cu-Zn ferrites. *Xinan Keji Daxue Xuebao*, 22(4), 10-13, 19.

Yang, G.; Ji, H.; Miao, X.; Hong, A.; Yan, Y. (2011). Crystal growth behavior of $LiFePO_4$ in microwave-assisted hydrothermal condition: from nanoparticle to nanosheet. *Journal of Nanoscience and Nanotechnology*, 11(6), 4781-4792.

Yang, Q.; Huang, J.-f.; Cao, Li-y.; Wang, B.; Wu, J.-p.; Yang, T. (2010). Preparation of mullite microcrystallites by sol-gel combined with microwave hydrothermal process. *Rengong Jingti Xuebao*, 39(6), 1456-1460.

Yang, Yi-F.; Ma, Y.-S.; Bao, S.-S.; Zheng, Li-M. (2007). Microwave-assisted hydrothermal syntheses of metal phosphonates with layered and framework structures. *Dalton Transactions*, (37), 4222-4226.

Yang, Yi-Lin; Hu, Chi-Chang; Hua, Chi-Chung. (2011). Preparation and characterization of nanocrystalline $Ti_xSn_{1-x}O_2$ solid solutions via a microwave-assisted hydrothermal synthesis process. Cryst. Eng. Comm., 13(19), 5638-5641.

Yuanyuan Hu; Chong Liu; Yahong Zhang; Nan Ren; Yi Tang. (2009). Microwave-assisted hydrothermal synthesis of nanozeolites with controllable size. *Microporous and Mesoporous Materials*, 119, 306–314.

Yoon, S.; Manthiram, A. (2011). Microwave-hydrothermal synthesis of $W_{0.4}Mo_{0.6}O_3$ and carbon-decorated WO_x-MoO_2 nanorod anodes for lithium ion batteries. *Journal of Materials Chemistry*, 21(12), 4082-4085.

Yu, J.C.; Hu Xianluo; Li Quan; Zheng Zhi; Xu Yeming. (2005). Synthesis and characterization of core-shell selenium/carbon colloids and hollow carbon capsules. *Chemistry*, 12(2), 548-52.

Yu, Y.-T.; Dutta, P. (2011). Synthesis of Au/SnO_2 core-shell structure nanoparticles by a microwave-assisted method and their optical properties. *Journal of Solid State Chemistry*, 184(2), 312-316.

Zawadzki, M. (2008). Microwave-assisted synthesis and characterization of ultrafine neodymium oxide particles. *Journal of Alloys and Compounds*, 451(1-2), 297-300.

Zeng, R.; Wang, J.-Q.; Chen, Z.-X.; Li, W.-X.; Dou, S.-X. (2011). The effects of size and orientation on magnetic properties and exchange bias in Co_3O_4 mesoporous nanowires. *Journal of Applied Physics*, 109(7), 07B520/1-07B520/3.

Zhang, L.; Cao, X.-F.; Ma, Y.-Li; Chen, X.-T.; Xue, Zi-L. (2010). Pancake-like $Fe_2(MoO_4)_3$ microstructures: microwave-assisted hydrothermal synthesis, magnetic and photocatalytic properties. New Journal of Chemistry, 34(9), 2027-2033.

Zhao, Q.; Yang, Y.; Sun, Y.-x.; Wang, S.-g.; Liu, L.; Chang, A.-m. (2007). Microwave hydrothermal synthesis of yttria stabilized zirconia at low temperature. *Weinadianzi Jishu*, 44(7/8), 76-79.

Zheng, Y.; Tan, G.; Bo, H.; Miao, H.; Chang, M.; Xia, A. (2011). Effect of Tween 80 on preparation of $BiFeO_3$ in microwave-hydrothermal method. *Guisuanyan Xuebao*, 39(8), 1249-1253.

Zheng, Ya-lin; Xu, G.-l.; Lai, Z.-yu; Liu, M. (2007). Effects of Cu content on the sintered properties of Ni-Cu-Zn ferrites. *Yadian Yu Shengguang*, 29(6), 707-709.

Zhou, Y.; Yu, J.; Guo, M.; Zhang, M. (2010). Microwave hydrothermal synthesis and piezoelectric properties investigation of $K_{0.5}Na_{0.5}NbO_3$ lead-free ceramics. *Ferroelectrics*, 404, 69-75.

Zhu, Z. F.; He, Z. L.; Li, J. Q.; Liu, D. G.; Wei, N. (2010). Synthesis and characterisation of fluorinated TiO_2 microspheres with novel structure by sonochemical-microwave hydrothermal treatment. *Materials Research Innovations*, 14(5), 426-430.

Chapter 6

Dechlorination of Polyvinyl Chloride in NaOH and NaOH/Ethylene Glycol Solution by Microwave Heating

F. Osada and T. Yoshioka

Additional information is available at the end of the chapter

1. Introduction

The amount of plastic produced in Japan was 14.65 million tons in fiscal year 2007, although this figure has fluctuated, as shown in Figure 1. As a result, the amount of plastic waste has increased to 10.05 million tons/year[1]. Most of that waste is either landfilled or incinerated.

The amount of plastic waste discharged is annually growing. Therefore, it is essential to research and develop technologies to recycle and reuse plastic waste to protect the global environment and effectively to utilize resources. In an effort to promote recycling of plastic waste. The Containers and Packaging Recycling Law was enacted in June 1995 to promote sorted collection of containers and packaging and the recycling of containers and packaging as products. To promote recycling, the scope of this law was expanded to plastic containers and packaging, instead of being limited to polyethylene terephthalate (PET) bottles. When it was amended then reenacted in April 2000, it seemed as if almost 50% of plastic waste was recycled in fiscal year 2000. However, the breakdown of effective utilization in fiscal year 2006 revealed that almost 72% was simply incinerated: [1] 40% was used for power generation, 19% was incineration to release thermal energy, 4% was used as feedstock, 9% was used as solid fuel (Figure 2). Some waste plastic from general household was also incinerated in the metropolitan area. Thus, the current situation is that most waste plastic is still either landfilled or incinerated. There is a limit to how much waste plastic that a landfill can hold, and the option of landfilling waste plastic is becoming more difficult, especially since the amount of landfill space available to bury plastic waste is annually decreasing. In fact, it has been reported that the landfills will be completely full in 14.8 years (as of end of fiscal year 2005) for general waste, and in only 7.7 years (as end of fiscal year 2005) for industrial waste. When plastic waste is landfilled, there is risk of endocrine disruptors and hazardous substances eluting from the waste. The specific gravity of plastic is low. Therefore, burying it poses risk of loosening the landfill ground. This makes it difficult to reuse the landfill site for other purposes. On the other hand, when waste plastic is incinerated, it tends to corrode the incinerator due to the waste plastic containing halogens such as chlorine. What is more, depending on the incineration condition, it can also generate toxic organic compounds and materials causing acid rain. Burned ash and fly ash that are buried in landfills also contain residual $CaCl_2$ from $Ca(OH)_2$ added to neutralize hydrogen chloride. This also increases the load on the landfill. Among waste plastic, polyvinyl chloride (PVC), whether rigid or flexible, is common because it is used in construction materials and materials for electrical wire sheathing. PVC is used as a durable and enduring product, thus, there is risk of an increase in volume discharged when buildings are renovated or rebuilt in the future.

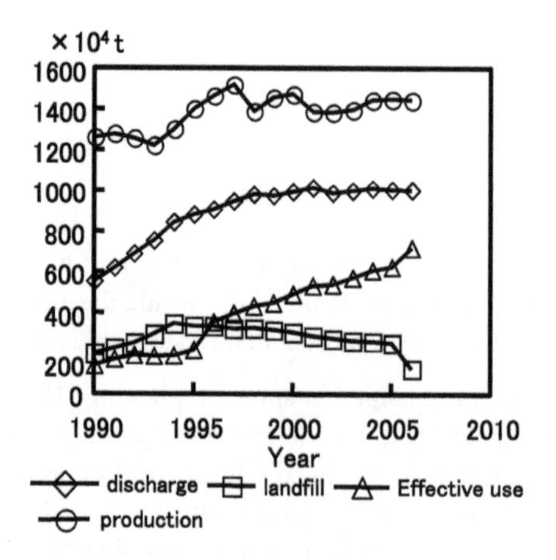

Figure 1. Production and emissions of plastics

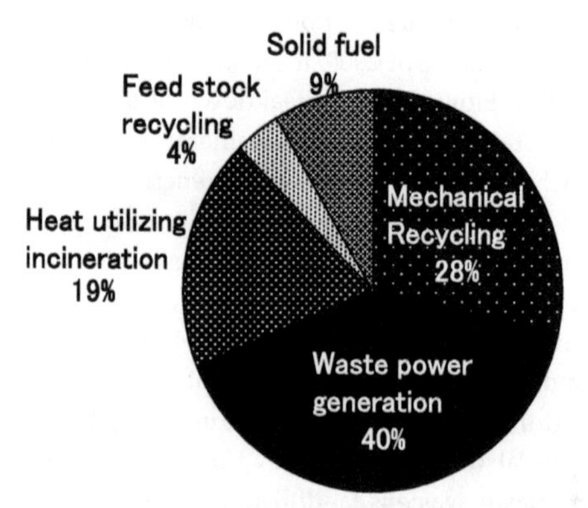

Figure 2. Breakdown of waste plastics to the effective use (2006)

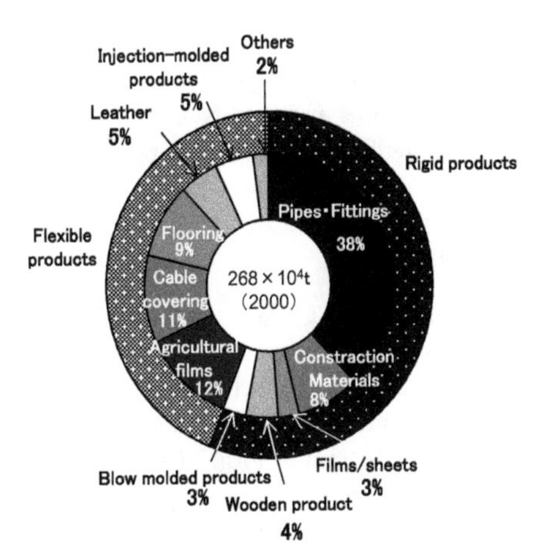

Figure 3. Applications of PVC

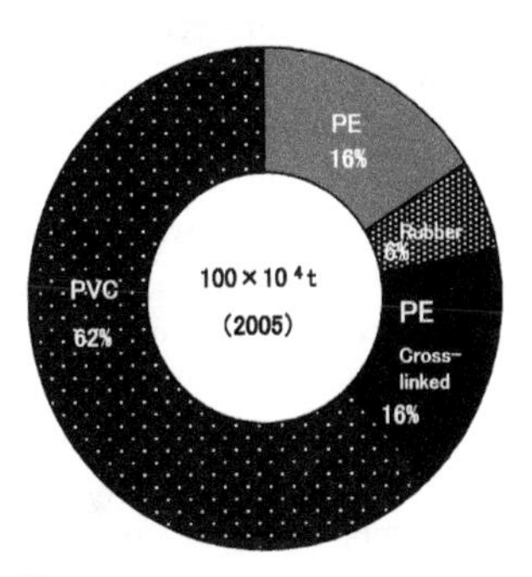

Figure 4. Waste material coated cables

There have been many studies on PVC since 1950. However, the objectives of these studies tended to mainly focus on stabilization. Recent reports indicate that more studies have focused on recycling as a study objective. Examples include:

1. Material recycling to make industrial raw materials,
2. Recycling to turn waste vinyl chloride products into new vinyl chloride products,
3. Chemical recycling (feedstock recycling) [2] by fractional precipitation,
4. Recycling PVC waste products into chemicals and raw materials for products through chemical treatment methods [3],
5. Energy recycling by thermal energy recovery from incineration of PVC wastes, dechlorinated via thermal degradation, and by power generation from incineration of PVC waste and gasified or liquefied PVC waste, reusing this energy source to generate power.

PVC studies are largely classified into two types: the dry method by thermal degradation and wet method. When considering the speed of temperature rise during thermal degradation of pure PVC powder, with the dry method we can get type-A crude oil with an average molecular weight of 280.[4]With the wet method, we can process flexible PVC. The plasticizer additive becomes anhydrous and then NaOH solution is added. PVC undergoes hydrolysis, and the heavy oil contents are pressurized and reheated to make light crude oil, thereby turning PVC into an oil.[5] NaOH solution can be used for dechlorination, then PVC is converted into carboxylic acids.[6] Flexible PVC film can be deplasticized and dechlorinated to synthesize polyene.[7]

We must find new methods to handle PVC waste using approaches that are not dependent on landfills and incineration. To do this, there is a need to create products with a very long usage life, as well as to promote reuse and recycling of products. If it is difficult to reuse or recycle these products, then we must support material recycling by recycling it into something else. Films are the most common form of flexible PVC,[8] as seen in the PVC product breakdown shown in Figure 3. It is most used in agricultural vinyls, as well as in packaging for agricultural fertilizers. Until recently, these items were incinerated. However, we see that tendency has changed, and that they are now often washed of adherents and residue and then reused/recycled or reused/recycled into lower quality PVC products. However, this amounts to only a fraction of the PVC waste discharged. Most of it is either incinerated or exported to Southeast Asia and China to reduce the treatment cost. The second largest waste comes from electrical wire sheathing. Roughly 1 million tons of cable sheathing is discharged as waste every year, of which 62% is PVC [9](Figure 4). About 300,000 tons, or roughly half, of this PVC undergoes material recycling and is made into flooring materials or other lower quality products, and a fraction is recycled into electrical wire sheathing materials. The reality is that the remainder is either landfilled or incinerated. The majority of PVC products discharged from different industrial fields is generally either landfilled or incinerated; however, some may be recycled. In either case, the situation is that the mode of disposal still tends to be mostly dependent on incineration. For waste that can only be incinerated, this study intends to find a way to extract and remove beforehand, thus eliminating the hydrogen chloride that would be generated upon incineration. A flexible PVC tube was used as the source of PVC in the study. We then studied whether use of external heating with a general thermal heater or internal heating with a microwave heater could promote the reaction conditions when used as the heating source for the deplasticization and dechlorination processes. It is generally known that polyvinyl chloride (PVC) is dehydrochlorinated when burnt at temperatures higher than 280°C. However, to effectively accelerate dehydrochlorination process with faster and better heat transport. Substances with low thermal conductivity are generally heated by microwaves because this procedure does not require heat transport. Thus new applications utilizing microwaves are expected to be developed in the near future. In this study, we reviewed the dehydrochlorination of flexible PVC to see if it was possible to use controllable microwave heating as a fast heating mechanism. This assumption was based on conventional wet method studies, [10-16] which indicated that substance containing Cl molecules tended to easily absorb microwaves, and materials made of flexible PVC and ethylene glycol (EG) used as a reacting solution also easily absorbed microwaves. The goal was also to accelerate dehydrochlorination by directly heating PVC molecules for early extraction of plasticizers.

We similarly studied the conditions that best promote dehydrochlorination and improve reaction efficiency through an interactive effect by using substances such as EG that present good intersolubility with flexible PVC.

PVC	59.2
DOP	29.7
Epoxidized soybean oil	3.6
Other substances	7.5
Cl	35.7

Table 1. Composition of flexible polyvinyl chloride (PVC) (wt%)

2. Experimental study

2.1. Sample and reagent

The materials used included flexible PVC with the composition indicated in Table 1. The weight ratio of dioctyl phthalate (DOP) plasticizer was roughly 30%, compared to a weight ratio of roughly 60% for PVC polymer. Other components included the oil added upon molding and UV stabilizers. We confirmed that the PVC polymer contained 35.7% of chlorine according to elementary analysis. We also used NaOH (Merck GR99%) diluted in distilled water as the NaOH solution for extraction and removal treatment in this study. Ethylene glycol was used without adding anything.

2.2. Reactor

A general purpose microwave heater (Milestone General) was used for the reaction experiments.[18] The heater had a fixed microwave frequency of 2.45 GHz, a maximum output of 1kW, a maximum usable temperature of 240°C, a maximum usable pressure of 3.4MPa, with a 100-ml reactor vessel. The reactor vessel was made of tetrafluoroethylene perfluoroalkylvinylether (PFA) copolymer, and the outer tube was made of polyether ether ketone (PEEK). In the container was also a magnetic stirrer so that the contents could be stirred. The mechanism also came with a temperature sensor to measure the temperature of the solution in the container. The temperature sensor measurement was also fed back to control microwave output. The inside of the device, including the reactor vessel, was not easily heated by microwaves, and both the reactor vessel and outer tube protecting the reactor vessel were made of materials that easily transmit microwaves. To ensure even heating distribution, the reactor vessel was also placed on a rotary stand. When heating was completed, the inside of the device and reactor vessels were cooled by a built-in fan. Conditions such as the reaction temperature and heating time were controlled using the exclusive control settings. The temperature elevation time, hold temperature, and temperature hold time can be input into the oven in advance, and the oven operates automatically.

2.3. Experimental method

In this study, we used a reactor vessel with a volume of 100ml and added 0.5g of flexible PVC crushed to a maximum particle size of 1mm and 50ml of NaOH solution. The solution was continuously agitated. The container containing the sample was set on the microwave heater and the sequence program was then started for microwave irradiation according to the preprogrammed target reaction temperature, heating time, temperature hold time, temperature elevation time and other parameters. The reaction start time was defined as the time when the solution in the vessel reached the set temperature. Then the temperature was maintained for the specified time. NaOH was prepared at concentrations of 2, 4, 6, 8 and 16mol/l, and the reaction temperature was set at 100°, 150°, 200°, 225° and 250°C. After the temperatures hold time had elapsed, agitation was stopped and the reaction products were cooled to room temperature. Then the solution was passed through a 0.45-mm polyvinylidene difluoride (PVDF) filter to separate the residue and filtrate, and then distilled water was added to the filtrate to prepare a 50-ml reaction solution. The residue was then washed with distilled water and dried by decompression; it was then weighted and underwent elementally analyzed. In the case of using NaOH/EG solution, the flexible PVC (Table 1.) with a particle size of about 1 mm was charged with 50 ml of EG were placed in the reactor vessel. NaOH was prepared at concentrations of 0.5, 1, 2, and 4 mol/l. The reaction temperature was set at 100, 120, 140, and 160°C.

3. Analysis

3.1. Chlorides and phthalic acid

The anion levels in the filtrate were measured using Ion Chromatograph (Model 7310-20, NIKKISO, Tokyo, Japan) consisting of DIONEX AG15 and AS15 columns. The eluent was mixed with NaOH solution of 35×10^{-3}mol/l, which was continuously added at 0.75ml/min. The ASRS suppressor was operated in recycle mode. The filtrate was then accordingly diluted with distilled water. The phthalic acid ions and chloride ions were quantitatively analyzed by concentrated analysis using concentrated columns.[19]

3.2. Residue

The residue recovered was washed with distilled water and methanol and then dried by decompression. After drying, the residue structure was examined under a scanning electron microscope (Hitachi S-2510). The change in molecular weight was then examined with a gel permeation chromatography (GPC) analyzer (Waters).

3.3. Evaluation method

The dechlorination ratio and deplasticization ratio were defined and evaluated as follows;

$$\text{Dechlorination ratio (\%)} = [Cl_t]/[Cl_0] \times 100 \tag{1}$$

$$\text{Deplasticization ratio (\%)} = [DOP_0]/[DOP_t] \times 100 \quad (2)$$

Where $[Cl_0]$ is the number of moles of chlorine in the original flexible PVC and

$[Cl_t]$ is the number of moles of chlorine in the in the extracted solution after reaction time t.
$[DOP_0]$ is the number of moles of plasticizer in the original flexible PVC and
$[DOP_t]$ is the number of moles of plasticizer in the extracted solution during after reaction time t.

	Thermal process	Frequency
Induction heating (heating of the metal)	Low frequency heating	50/60Hz
	High frequency heating	Several hundred KHz
Dielectric heating (heating of the insulator)	Dielectric heating	1~400KHz
	Microwave heating	0.3~300GHz

Table 2. Thermal processes and frequency

4. Results and discussion

An electric heater is generally used as the heat source of an autoclave. The external heating method heats the surface of an object with a heat source. With this method, the object is gradually heated from the outside to the inside according to the thermal conductivity of that object. In other words, heat is transferred from the surface in which the treatment vessel and electric heater come in contact. The heat is then sequentially propagated from the liquid to the object being heated. In contrast, internal heating by high-frequency dielectric heating by electromagnetic waves or microwave heating are intense heating methods where objects are heated from the inside then outward. This tends to be a highly efficient heating method since there is practically no heat escaping outside; the system delivers swift item heating, as we know from what a domestic microwave oven can do. Polarization takes place when a dielectric is exposed to an electric field, since charged carriers (electrons and ions) in a dielectric move around, while the dipoles try to face the direction of the electric field. Heating method and their associated frequency ranges are illustrated in Table 2.[20] In microwave electric fields, the polarity reverses 2.45 billion times per second. Heat is generated through intense rotation, collision, vibration and friction of the dipoles influenced by the oscillating electric field.[21] This exothermic reaction is expressed in Eq.3, showing the amount of electric power (P) consumed per unit area in the dielectric;

$$P = (E/d)^2 \times 5.56 \times 10^{-11} \times f \times \varepsilon r \times \tan\delta (W/m^3) \quad (3)$$

Where E is the intensity of the electric field (V), d is the distance between electrodes (m), f is the frequency (Hz), εr is the relative permittivity, $\tan\delta$ is the dielectric tangent, and P is the electric power consumption per cubic meter (W/m^3).

Material	Tangent*δ (× 10)
Water	1570.0
Sodium Chloride	2400.0
Ethylene glycol	10000.0
Fused Quartz	0.6
Ceramic	5.5
Nylon66	128.0
Polyvinyl Chloride	55.0
Polyethlene	3.1
TeflonPFA	1.5

Table 3. Material loss factor (at 3GHz)

The product of εr and tanδ is called as loss coefficient, which is specific to each material.[20] and also changes with the temperature and frequency. According to Eq.3, we see that the level of heating energy increases when the intensity of the electric field, the frequency, and the loss coefficient increase.

PVC shows asymmetries in polarization due to its molecular structure. Therefore, it is classified as a plastic with a polarity. Other examples include acrylonitrile butadiene styrene (ABS) resin, nylon, and polyvinyl acetate. As indicated in Table 3, plastic with a polarity have a larger loss coefficient; therefore, they easily absorb microwaves.[22] On the other hand, plastic without polarity, such as Teflon, polyethylene, and polypropylene, have a smaller loss coefficient; therefore, microwaves easily pass through them. For this reason, Teflon is used as a structural material when using microwaves, or as a heat- and chemical-resistant container for chemical experiments using microwaves. Polyethylene and polypropylene are used as containers for preserved food and meals that are to be heated, with regard to their heat- resistant properties. On the other hand, plasticizer is mixed with and coexists with flexible PVC. It does not form a chemical bond with PVC. Significant vibration and heat is generated between the molecules since plasticizer is mixed/coexists with PVC and because plasticizer itself absorbs microwaves.

As a result, the plasticizer and PVC are easier to separate.

It was discovered that deplasticization and dechlorination could take place separately by changing the reaction temperature. The purpose of this study was to separately isolate and recover extracted and chlorides in the same reaction solution in a way that the plasticizer and chlorides do not coexist, and to do this to the greatest extent possible by setting the reaction temperature at two different levels. We detected chloride ions, phthalate ions, isooctanol and hexanol in the solution extracted after the reaction.[7] The residue was yellow, brown, or black and was smaller in volume than the initial flexible PVC sample. It has been reported that when there are more than seven conjugated double bonds created by

PVC dechlorination, they absorb visible light, which makes dechlorinated PVC appear yellow, brown or black.[21] Under the initial temperature conditions to solely extract plasticizers, the plasticizer was hydrolyzed and extracted into phthalic acid and isooctanol, then extracted as sodium phthalate in the solution (Eq.4). At this point, no chloride ions were detected. Thus, it is conjectured that dechlorination has not yet taken place. Next, we detected chloride ions by raising the reaction temperature to the temperature condition to only perform the second level of dechlorination. Based on the detection of chloride ions, it was conjectured that PVC dechlorination selectively occurred according to the reaction temperature (Eps.5, 6). From the above findings, we found that it was possible to both extract plasticizer and dechlorinate PVC in high-temperature alkaline solution using microwave heating (internal heating) by simply changing the reaction temperature, as illustrated in the reaction process of Scheme 1.[7]

First reaction

$$C_6H_4(COOCH_2CH(C_2H_5)C_4H_9)_2 + 2\,NaOH \xrightarrow[\text{hydrolysis}]{} C_6H_4(COONa)_2 + 2HOCH_2CH(C_2H_5)C_4H_9 \quad (4)$$

Second reaction

E2

$$-\!\left(CH_2\text{-}CH(Cl)\right)_n\!- + NaOH \longrightarrow HO^-:\ -\!\left(C(H)(H)\text{-}C(H)(Cl)\right)_n\!- \quad (5)$$

Elimination reaction

$$\longrightarrow -\!\left(CH=CH\right)_n\!- + NaCl$$

Sn2

$$-\!\left(CH_2\text{-}CH(Cl)\right)_n\!- + NaOH \longrightarrow -\!\left(CH_2\text{-}CH(OH)\right)_n\!- + NaCl \quad (6)$$

OH^-

Nucleophilic Substitution

When using conventional external heating sources, we found that both the extracted and removed chlorides coexisted in the reaction solution after deplasticizing and dechlorinating flexible PVC.[7,24] With our microwave-heating method, we found that it was possible to extract plasticizer and remove chlorides at different temperatures. First the plasticizer is extracted from NaOH solution. Then the same NaOH solution is heated at a higher reaction temperature to remove the chlorides. By using this method, we successfully removed the chlorides from PVC without mixing them with any plasticizer. We speculate that the reactions are dependent on reaction temperature. As a result, we can say that internal heating using microwaves allows us to selectively and individually extract plasticizer and remove chlorides one after the other by simply changing the reaction temperature, which is not possible when using external heating sources such as conventional electric heaters. In the case of using the NaOH/EG solution, plasticizer that is mixed with flexible PVC will be easy to heat since EG (with high loss coefficient) can easily permeate it. Since the plasticizer is not chemically bonded with PVC, exposure to microwave radiation generates, powerful motion between molecules and generates heat. By separately heating the plasticizer and PVC, and by setting the reaction temperature at two different levels, we assume that it becomes easier to separate and extract the plasticizer.

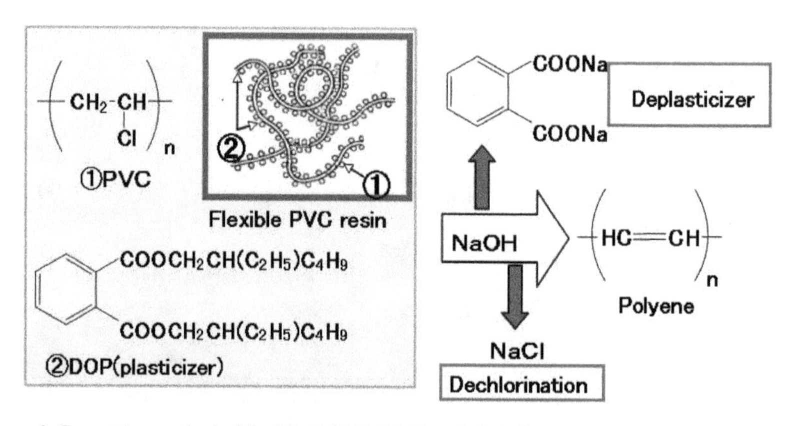

Scheme 1. Processing method of flexible PVC, DOP, dioctyl phthalate

4.1. Dehydrochlorination

Weight loss was measured when using 8mol/l NaOH solution at a reaction temperature between 150° and 250°C for a reaction time of 30min. The measured values are shown in Figure 5. The weight loss ratio was 24.4% at 150°C, 34.8% at 200°C, 62.5% at 225°C, and 71% at 250°C. Since the DOP plasticizer content was 29.7% of the total compound weight, 82% of DOP was hydrolyzed at 150°C and almost all was hydrolyzed and extracted at 200°C. What is more, since weight loss reached 71% at 250°C, this further change in weight indicates the removal of additives other than DOP and, mainly, the dechlorination of PVC. These results indicate that it is possible to selectively extract plasticizer and remove chlorides in a NaOH solution by changing the reaction temperature to achieve deplasticization and dechlorination. This suggests the possibility of inhibiting the generation of organic chloride compounds by extracting the plasticizer beforehand from the flexible PVC. The results of the dechlorination of flexible PVC tube using 2-16mol/l NaOH solution at a reaction temperature of 150°-250°C are shown in Figure 6. Our results indicate that the concentration of NaOH solution did not affect dechlorination, but the reaction temperature did greatly affect dechlorination. Flexible PVC has two levels of degradation that are dependent on reaction temperature. Below 200°C, hydrolysis and extraction of DOP mainly took place, but dechlorination did not.

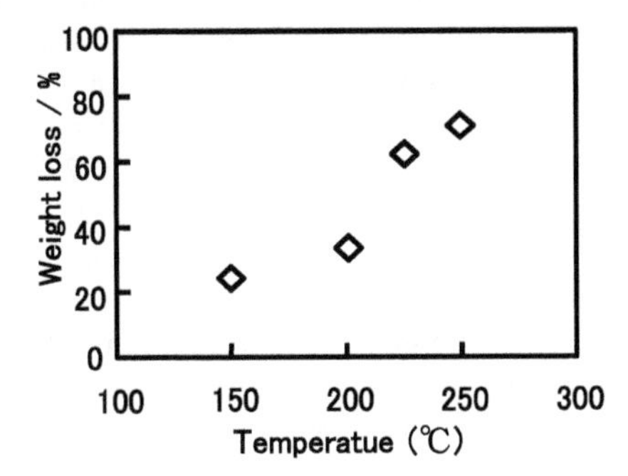

Figure 5. Effect of temperature on the weight loss in 8mol/L NaOH for 30 min by microwave heating

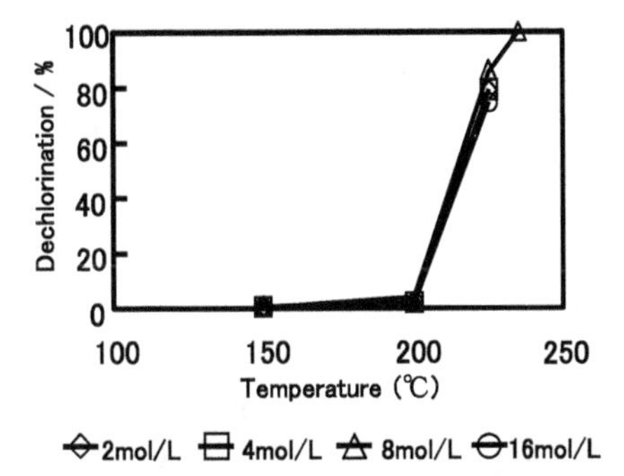

Figure 6. Dechlorination as a function of temperature and NaOH concentration for 30 min by microwave heating

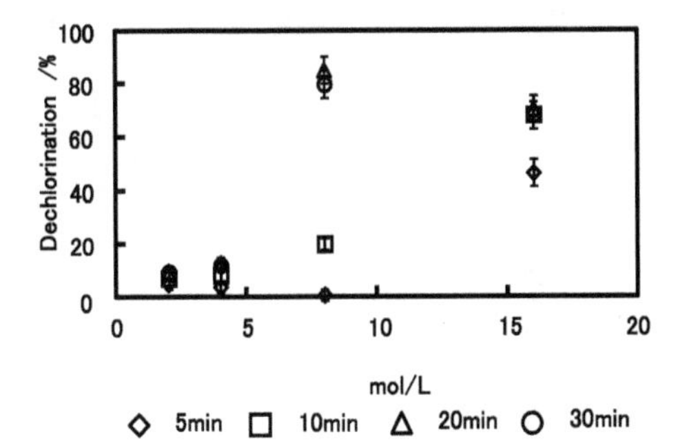

Figure 7. Dechlorination of flexible PVC as a function of NaOH concentration and reaction time at 225°C with microwave heating

It was assumed that once the DOP plasticizer was completely hydrolyzed and extracted, the microwaves rapidly promoted dechlorination, since the hydrolyzed DOP was in a state in which it could be absorbed by PVC chlorides. The results of dechlorination using 2-16mol/l NaOH solution at reaction temperature of 225°C for various reaction time are shown on Figure 7. In 2 and 4mol/l NaOH solution, there was no significant increase in dechlorination ratio, even when the reaction time was increased. In 8mol/l NaOH solution, the dechlorination ratio was 0% at 5min, 20% at 10min, 85% at 20min, and 80% at 30min, showing almost constant progress. In 16mol/l NaOH solution, the dechlorination ration reached 45% at 5min., however, the ration remained constant at 70% although the reaction time was extended. After DOP plasticizer was hydrolyzed and extracted, the tendency was for the flexible PVC sample to became smaller and decrease in surface area, as shown in the surfaces presented in Figure 8a-e. At 150°C, when DOP plasticizer was hydrolyzed and phthalic acid was extracted, it created micro-pores measuring roughly 3μm in diameter size (Figure 8b). The micro-pores grew in size as the reaction temperature was elevated. According to Figure 8d, micro-pores were formed of about 1-2μm in size overall as a result of rapid progress of dechlorination. It was also found that the sample reduced in size compared to the pre-reaction state of Figure 8a. Based on these states, the contact area with the NaOH solution was no longer sufficient. As a result, although the reaction time was extended, it resulted in delayed progress of dechlorination. Further, it may be considered that the dechlorination ratio did not improve or was not complete because the microwave output for internal heating was automatically controlled according to the liquid temperature of the NaOH solution.

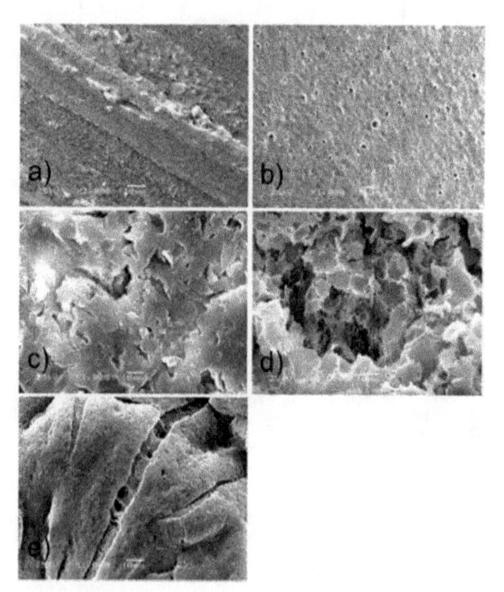

Figure 8. Scanning electron microscope images of residues from treatment of PVC with 8mol/l NaOH: a before reaction, b at 150°C, c at 200°C, d at 225°C, and e at 235°C

Table 4 illustrates the elemental analysis of the solid residue obtained after reactions in 8mol/l NaOH solution. As can be seen, the chlorine ratio was higher in the residue at 150°C than the sample that was used in our experiment. Based on the results described earlier, we know that the deplasticization ratio is 100% at this temperature. It is speculated that this is mainly due to hydrolysis of DOP plasticizer and extraction of phthalic acid. It is also obvious that the amount of chlorine in the residue decreased as the reaction temperature was elevated above 150°C. Although H/C is 1.62 in the residue formed at a reaction temperature of 150°C, this ratio tended to decrease as the reaction temperature was elevated, and H/C was 1.05 at 235°C a temperature at which 100% DOP plasticizer extraction and dechlorination is achieved. After dechlorination, H/C was close to 1.0, the level at polyene is generated. Therefore, this is indicative of the solid residue containing 96% polyene. It is conjectured that the remaining 4% was a result of the reaction substituting the OH group. When carrying out the reaction for the same time with external heating instead of microwave heating, we obtained the results shown in Figure 9. At a reaction temperature of 250°C when the plasticizer was 100% removed, 20% dechlorination had already occurred. This means that the reaction solution contained coexistence phthalic acid extracted from the plasticizer and chloride ions at the same time. This makes it difficult to separate them later. However, with microwave heating, there is only the need to change the reaction temperature to allow the plasticizer and chlorides to be recovered separately. On the other hand, in the case of NaOH/EG solution, we studied the weight loss of flexible PVC for NaOH concentrations in NaOH/EG solution of 0.5,1, and 2mol/l at reaction temperature between 100 and 160°C and a reaction time of 10 min. A maximum weight loss of 74.7 % was identified at 160°C. There was no difference in dehydrochlorination results up to 120°C, although the NaOH concentration was modified, and dehydrochlorination was about 8% in a 1mol/l NaOH/EG solution. The dehydrochlorination rate increased in the 1mol/l NaOH/EG solution when the reaction temperature rose above 130°C and reached a maximum rate99.7% at160°C (Figure 10). We presume that the DOP hydrolysis reaction mainly occurred up to 130°C, and suggest that the PVC dehydrochlorination reaction mainly occurred at temperature above 130°C. Based on these results, we found that dehydrochlorination was possible at temperature that were 50°C lower than conventional external heating systems. This indicates that EG permeates the PVC particle, and when directly heated, the EG loss coefficient is large, thereby making it easy to absorb microwaves. The NaCl created around the PVC particles further improved microwave absorption (Table 3), showing reaction-promoting effects. This potentiation supported a rapid start to dehydrochlorination and efficient progress of the procedure (Table 3).[11]

	Content (wt%)			Atomic ratio	
	C	H	Cl	H/C	Cl/C
PVC*	51.6	6.8	35.7	1.56	0.23
150℃	42.3	5.8	48.8	1.62	0.4
225℃	84.4	7.8	6.2	1.1	0.03
235℃	88.1	7.8	N.D.**	1.05	-

*PVC : raw material, **N.D. : not detected

Table 4. Elemental analysis of residue formed at different temperatures

4.2. Removal of the plasticizer

The DOP plasticizer that present in flexible PVC is hydrolyzed in high-temperature alkaline solutions yielding phthalic acid and isooctanol that can be extracted from the solution. We studied the optimal reaction conditions to extract phthalic acid from DOP plasticizer.

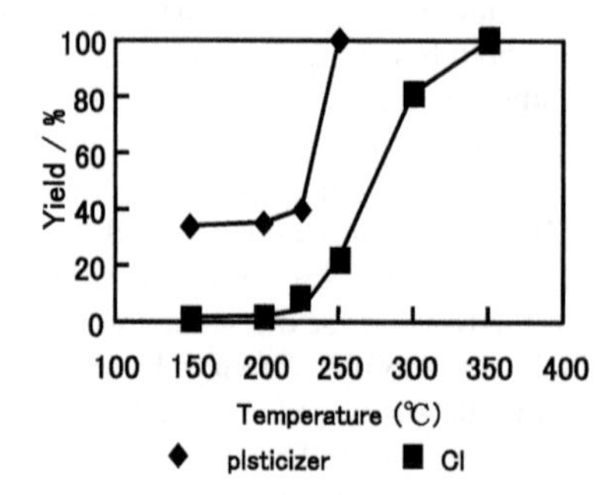

Figure 9. Effect of temperature on the yield of phthalic acid and Cl in 16mol/L NaOH solution using conventional heating

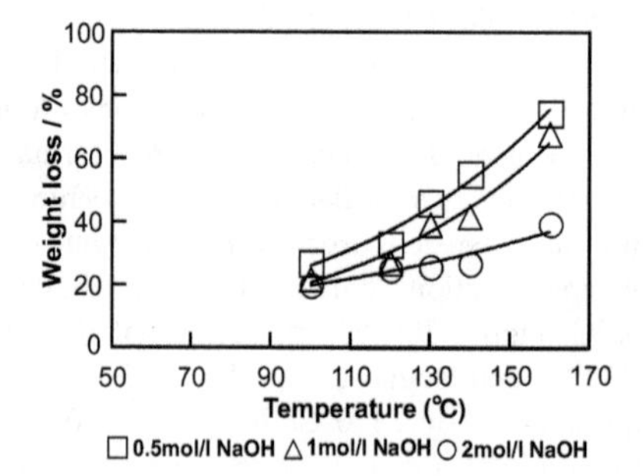

Figure 10. Effect of NaOH concentration on the weight loss from flexible PVC due to dechlorination and deplasticization on microwave heating for 10 min. Concentration of NaOH: squares, 0.5 mol/l; triangles, 1mol/l; circles, 2mol/l

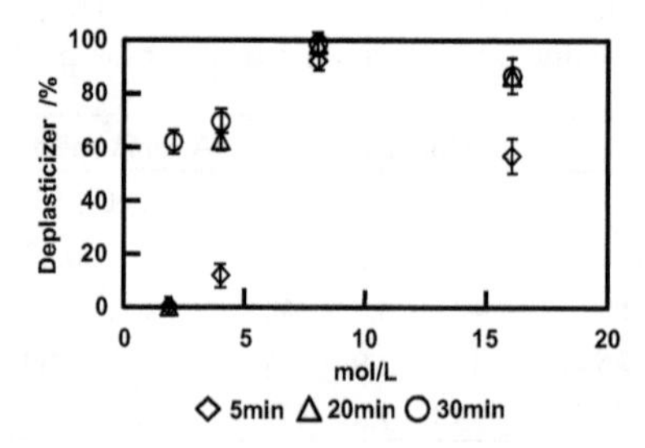

Figure 11. Deplasticization of flexible PVC as a function of NaOH concentration and reaction time at 150°C with microwave heating

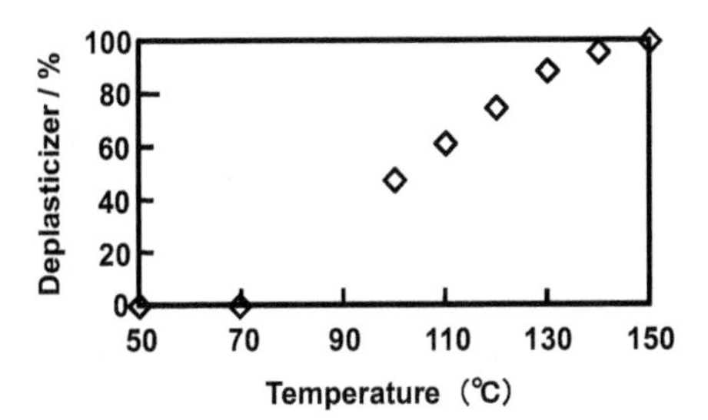

Figure 12. Deplasticization of flexible PVC as a function of temperature in 8mol/L NaOH for 30 min

We investigated the effects of NaOH solution concentration on the deplasticization ratio by microwave heating at a reaction temperature of 150°C for 30min (Figure 11). As can be seen, the recovery of phthalic acid from DOP plasticizer tended to be elevated at higher NaOH concentrations. The maximum recovery ratio was 63% in 2mol/l NaOH solution, 70 % in 4mol/l NaOH solution, 99% in 8mol/l NaOH solution, and 87% in 16mol/l NaOH solution. In 8mol/l NaOH, the recovery ratio was optimum. There was a tendency for the deplasticization ratio to drop when the concentration of NaOH solution increased further. As indicated in Figure 8b-e, when the plasticizer was eluted, the PVC sample became smaller. As a result, it was presumed that there was not enough contact with the NaOH solution at a high concentration. For this reason, there was not sufficient hydrolysis and the deplasticization ratio dropped as a result.

We studied the effects of reaction temperature on recovery ratio of phthalic acid in 8mol/l NaOH solution (Figure 12). The recovery ratio of phthalic acid was 0% at 50- 70°C. However, the recovery ratio of phthalic acid drastically increased when the reaction temperature exceeded 100°C, and reached 50% at 100°C and 100% at 150°C. This indicates that DOP plasticizer was hydrolyzed and extracted as sodium phthalate by microwave heating, since DOP plasticizer absorbed microwaves and effectively reacted with the NaOH solution.

Before the chemical reaction, the flexible PVC tube was translucent. At 70°C or lower, the color was the same as the raw material. However, at 100°C it turned light cream in color, and darkened as the temperature was elevated, finally turning orange at 150°C. This color change is speculated to be the result of an increase in double bonds due to some chlorides slightly escaping from the side chain of the PVC polymer.

The GPC results for molecular weight analysis of flexible PVC after treatment are shown in Table 5. Mn is the number-average molecular weight and Mw the weight-average molecular weight, expressed as shown inEq.7, when the number of molecules of component i with molecular weight Mi of a sample is N (i=1, 2, … q):

$$M_n=\sum_{i=1}^{q} M_iN_i/\sum_{i=1}^{q} N_i \quad M_w=\sum_{i=1} M_i^2N_i/\sum_{i=1} M_iN_i \tag{7}$$

Mn and Mw showed a declining tendency with increasing temperature. This seems to indicate that the molecular weight decreased when the principal PVC chain broke. However, we believe there is a possibility that the residues after deplasticization can be recycled by mixing it with virgin PVC materials. In either case, there is a need for more detailed evaluation because this process results in product with a darker pigmentation. In the case of NaOH/EG solution, the effect was studied of NaOH concentration and reaction temperature on the extraction of plasticizer in NaOH/EG solution in the range 1-4mol/l at reaction temperatures between 100° and 160°C and a reaction time of 10 min. As a result, we found that the maximum plasticizer extraction effect was attained in 1mol/l NaOH/EG solution at a reaction temperature of 140°C and with a reaction time of 10 min. However, plasticizer extraction tended to drop as the NaOH concentration increased (Figure 13). This is presumed to be caused by the high NaOH surface tension making it difficult for the plasticizer and NaOH to sufficiently come into contact and thereby making it difficult to extract, or because DOP was assessed in terms of phthalate ions (and with the promotion of hydrolysis it might become impossible to detect phthalate ions). Figure 14 indicates the rate of extraction of plasticizer in 1mol/l NaOH/EG solution for a reaction time of 10 min and different reaction temperature. The highest extraction rate was 98 % in 1mol/l NaOH/EG solution at a reaction temperature of 160°C and a reaction time of 10 min (Figure 14). It is also indicated that the 1mol/l NaOH/EG for a reaction time of 10 minutes (Figure 14). It is indicated that the plasticizer hydrolyzed to form phthalic acid on microwave heating.

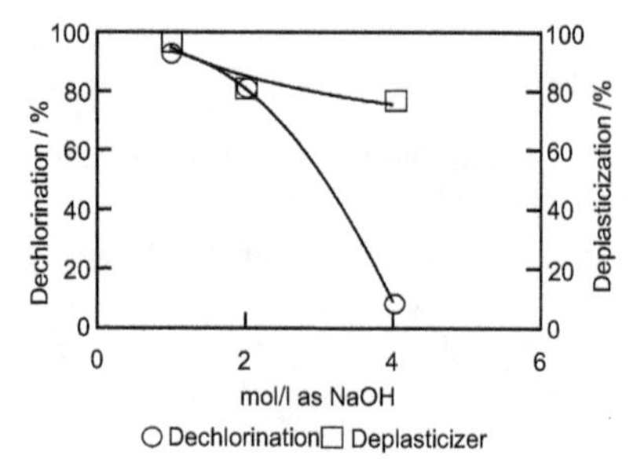

Figure 13. Extraction of plasticizer (*squares*) and dechlorination (*circles*) as a function of NaOH concentration using microwave heating at 190°C for 10min

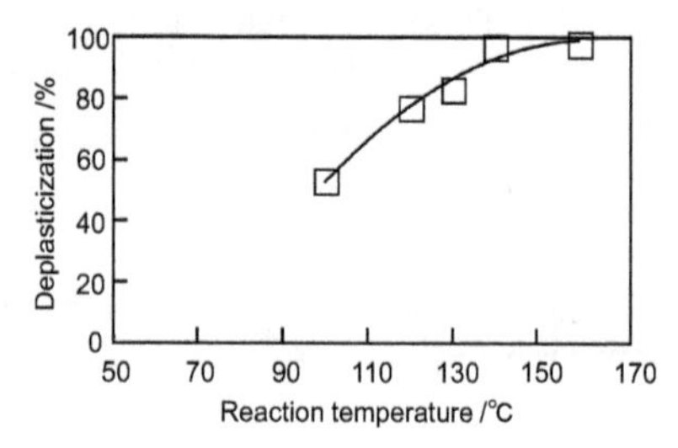

Figure 14. Extraction of plasticizer as a function of the reaction temperature in 1mol/l NaOH/EG using microwave heating for 10 min

	Mn	Mw
Before the reaction	114K	210K
100° C	111K	205K
110° C	108K	205K
120° C	109K	204K
130° C	108K	202K
140° C	104K	202K
150° C	96K	190K

Mn, number average molecular weight; Mw, weight average molecular weight $K=10^3$

Table 5. Molecular weight determination results from gel permeation chromatography (GPC)

4.3. Effect of NaCl in promoting heating

Let us consider water to illustrate the dynamics of the loss coefficient. When water gets hotter, the loss coefficient gets smaller (Table 6), which indicates lower microwave absorption. When looking at the NaOH or NaOH/EG solution and flexible PVC composition in the reactor vessel, we find that the constituents, NaOH, EG, PVC, and plasticizer, have different abilities to absorb microwaves. When the reaction solution and reagent temperature rises, making it more difficult to absorb microwaves their respective potentiating effects will promote extraction of the plasticizer. What is more, the HCl released by PVC dehydrochlorination reacts with NaOH to become NaCl, As a result, the loss coefficient increases; the NaCl created around the PVC absorbs microwaves and promotes the rise in inside temperature. This makes it possible to effectively dehydrochlorination PVC.

Temperature (°C)	Tangent*δ (× 10)
15.0	2050
25.0	1570
35.0	1270
45.0	1060
55.0	890
65.0	765
75.0	660
85.0	547
95.0	470

Table 6. Effect of temperature on the dissipation factor of water at 3GHz

5. Conclusion

In this study, we investigated how flexible PVC decomposed in a high-temperature alkaline solution heated by microwave to learn about the hydrolysis and extraction of DOP plasticizer and dechlorination process. We came to the following conclusion as a result of the study.

1. We found a way to selectively hydrolyze/extract DOP plasticizer and remove chlorides using high-temperature alkaline solutions by internally heating flexible PVC (made of defined sample components) using microwave heating.
2. The suitability of microwave heating was acknowledged. It was also possible to separately deplasticize and dechlorinate the sample by changing the reaction temperature or the reaction control temperature. When external heating was applied using a general electric heater (Figure 9), plasticizer and chlorides were found to coexist in solution since chloride removal takes place at 250°C, the reaction temperature required to extract 100% of the plasticizer. This requires an additional step to individually separate the plasticizer and chlorides. However, using microwave heating, it is possible to separately extract plasticizer and remove chlorides by simply controlling or changing the reaction temperature without the plasticizer and chlorides coexisting in the reaction solution.
3. We believe that it is possible to use microwave heating to extract 100% of the plasticizer with no dechlorination occurring by changing/controlling the reaction temperature. The residues in this case can be reused as materials recycled into PVC.
4. According to studies using external heating, the reaction temperature for deplasticization was 250°C and for dechlorination it was 350°C, using the same flexible PVC samples in a 16mol/l NaOH solution and using external heating with an autoclave with an electric heater. However, by using microwave heating as presented in this study, 100% deplasticization is possible at 150°C and 100% dechlorination is possible at 235°C in a 8mol/l NaOH solution, which was only half the concentration required when using an external heating sources.
5. It was found that microwave heating with NaOH/EG was suitable for flexible PVC dehydrochlorination and extraction of plasticizer.
6. Using an NaOH/EG solution, it was found that microwave heating possibly allowed the reaction temperature to be lowered from 190°C to 160°C, and shortened the reaction time from 60 to 10 min.
7. When microwave heating was used as a heating source and NaOH/EG solution was used, it was possible to lower the concentration of NaOH to 1 mol/l instead of the conventional 8 mol/l solution used to remove the chlorides from PVC. Further, the reaction time was also shortened from the conventional 30 min to 10 min, and thus, this procedure is considered as energy saving method.

Author details

F. Osada

Analysis Solution Engineering Section,
2nd Engineering Department Higashimurayama Plant Industrial Division, NIKKISO, LTD., Japan

T. Yoshioka

Graduate School of Environmental Studies, Tohoku University

6. References

Online journal

Plastic Waste Management Institute (2007) (See Japanese web site: http://www2.pwmi.or.jp/siryo/flow/flow-pdf/flow2007.pdf)

Journal

Hashimoto H, Kudo H, Ushiku M (1944) Proceedings of the 5th Annual Conference of The Japan Society of Waste Management Experts (JSWME) 1944:159-161

Shin S.M, Watanabe S, Yoshioka T, Okuwaki A (1997) Dechlorination behavior in high-temperature alkali aqueous solution in farm PVC system polymer film. Chemical Society of Japan 1997, vol.1:64–67

Tsuji T, Ikemoto H, Irita S, Sakai N, Shibata T, Uemaki O, Itoh H(1998) Thermal and Catalytic Degradation of Polyvinyl Chloride, The Chemical Society of Japan, No.8, p.546-550

Maezawa Y(1994), Toshiba review, 49, 5, p.343

Yoshioka T, Okuwaki A(1994), Chemical Industry, 7,P.43-50

Shin S. M, Yoshioka T, Okuwaki A (1998) The Behavior of Degradation and Characteristics of Produced Char in Agricultural PVC Films in Aqueous Solutions at Elevated Temperatures (in Japanese), Japan Society of Waste Management Experts 9:141-148.

The Japan Plastic Industry Federration(2007) (See Japanese web site: http://www.jpif.gr.jp/3toukei/conts/2007/2007-seihin-c.htm)

Murata K, Aiba K, Ooya S, Tominaga Y, Matsumoto T, Mizuno K, Motomiya H(September 2002) Development of Insulated Wire and Cable Using Recycled PVC, Furukawa Review, No. 22, pp.1-5

Shin S. M, Watanabe S, Yoshioka T, Okuwaki A (1997) Dechlorination behavior in high temperature alkali aqueous solution in farm PVC system polymer film, Chemical Society of Japan No1: 64-67

Shin S. M, Yoshioka T, Okuwaki A(1998) Dehydrochlorination behavior of flexible PVC pellets in NaOH solution at elevated temperature. J Appl Polym Sci 67:2171-2177

Shin S. M, Yoshioka T, Okuwaki A(1998) Dehydroc hlorination behavior of rigid PVC pellets in NaOH solution at elevated temperature. Polym Degrad Stab 61:349-353

Yoshioka T, Furukawa K, Sato S, Okuwaki A (1998) Chemical recycling of flexible PVC by oxygen oxidation in NaOH solution at elevated temperature. J Appl Polym Sci 70:129-135

Yoshioka T, Furukawa K, Okuwaki A (2000) Chemical recycling of rigid PVC by oxygen oxidation in NaOH solution at elevated temperature.

Wachi S, Yoshioka T, Mizoguchi T, Okuwaki A, (2004) A published proceedings: Written only in Japanese. Pp 83-84. The 7th FSRJ Symp

Yoshii N, Osada F, Yana J (2004) Recycling Process for Waste Plastic. Executive Committee, International Symposium on Innovative Research Fields, preprints: 163-172

Kim DS, Chang J-S, Kim WY, Kim HY, Park S-E (1999) Bull Korean Chem Soc 20:408-410

Osada F, Nagai K (2008) Converting automobile shredder residue into desified refuse-derived fuel (in Japanese). Jpn Soc Waste Manag Experts 19(5): 303-309

Osada F (1996) Trace ion Analysis of Ion Impurities on Secondary System of PWR Nuclear Power Plants, Thermal and Nuclear Power Engineering Society Shikoku Branch, p.21-30

Book chapter

Neas E. D,Collins M. J (1988) Introduction to microwave Sample Preparation,p7-32

Osada F, Nagai K (2010) Dehydrochlorination of polyvinyl chloride using amine additives (in Japanese). Jpn Soc Waste Manag Experts 21:1 pp19-29

Book chapter

H. M. Kingston,L. B. Jassie (1988) Introduction to microwave Sample Preparation,p93-154

Journal

Hirayama K (1954), Chemical Society of Japan No 75:667.

Shin S. M, Watanabe.S, Yoshioka.T, Okuwaki A (1997) Dechlorination behavior in high temperature alkali aqueous solution in farm PVC system polymer film, Chemical Society of Japan No1: 64-67.

Section 3

Applications

Chapter 7

Microwave Dielectric Heating of Fluids in Microfluidic Devices

Mulpuri V. Rao, Jayna J. Shah, Jon Geist and Michael Gaitan

Additional information is available at the end of the chapter

1. Introduction

The science of designing, manufacturing and formulating processes involving fluidic devices having dimensions down to micrometers is known as microfluidics. In the last two decades, in the areas of chemical and biochemical sciences, there has been a great interest towards using microfluidic systems, which are popularly known as micro-total analysis systems (μ-TAS) (Manz et al., 2010) or lab-on-a-chip systems. These systems improve analytical performance and also facilitate incorporating various functions of distributed systems on a single-chip instead of having a separate device for each function (Reyes et al., 2002). Temperature control inside microfluidic cells is often required in a variety of on-chip applications for enhanced results, without significantly affecting the temperatures of other building blocks of the μ-TAS. This is a big challenge because the microfluidic devices on the chip need to be selectively heated.

2. Need for heating microfluidic devices

Heat energy has been in use for stimulating (increasing) chemical and biochemical reactions (reaction rates), which otherwise proceed slowly under ambient conditions. Rapid, selective, and uniform heating of fluid volumes ranging from a few microliters to as low as a few nanoliters is vital for a wide range of microfluidic applications. For example, DNA amplification by polymerase chain reaction (PCR) is critically dependent on rapid and precise thermo-cycling of reagents at three different temperatures between 50 °C and 95 °C. Another important and related application, temperature induced cell lysing, necessitates fluid temperature in the vicinity of 94 °C. Other potential applications of heating in a microchip format include organic/inorganic chemical synthesis (Tu., 2011), the investigation of reaction kinetics, and biological studies, to name a few.

A number of conduction-based heating approaches have been reported for microfluidic systems that include embedded resistive heaters, peltier elements, or joule heating under electro-osmotic and pressure driven flow conditions. Generally speaking, these methods require physical contact or close proximity between a fluid and a heated surface to transfer heat from that surface to the fluid. In microfluidic devices, when the fluid volumes approach nanoliters, heating rates will be potentially limited by the added thermal mass of the substrates used for heat transfer, and not by the fluid volume. The transfer of heat in such manner can also result in heating of large, undesired substrate areas creating spatial limitations for integration of multiple analysis functions on a single substrate. Additionally, the implementation of these heating methods will be limited to high thermal diffusivity substrates, such as silicon and glass, to maximize heat transfer rates. However, such substrates due to their high cost and complexity of the fabrication process are unsuitable for use in disposable devices. Due to a number of inherent problems associated with contact-mediated temperature cycling, a number of research groups have focused on the development of non-contact heating approaches. These non-contact heating approaches include heating based on hot air cycling, heating based on IR light, laser-mediated heating, halogen lamp-based heating, induction heating, and heating based on microwave irradiation. Hot air based heating method utilizes rapidly switching air streams of the desired temperature and transfer of air onto either polypropylene tubes or glass capillaries. However, the control and application of hot air streams on micro-fabricated integrated systems may not be easily accomplished without an impact on other structures or reactions to be executed on the chip. An inexpensive tungsten lamp as an IR source for rapidly heating small volumes of solution in a microchip format can potentially limit the heating efficiency when applied to microchips with smaller cross-section because the tungsten lamp is a non-coherent and non-focused light source leading to a relatively large focus projection. Other light based heating methods have been demonstrated for microfluidic heating, but such systems, generally speaking, require lenses and filters to eliminate wavelengths that could interfere with the reaction, and accurate positioning of the reaction mixture at the appropriate focal distance from the lamp, which further complicates their implementation.

Microwave dielectric heating is a candidate to address these issues. Most chemical and biochemical species are mainly comprised of water or solvated in water. Water is a very good absorber of microwave energy in the frequency range 0.3 GHz – 300 GHz. Due to this reason, microwave dielectric heating has been exploited for over five decades for heating and cooking fluids and food items containing water molecules (Brodie., 2011). It is also a very good candidate for implementing the heating function in chemical and biochemical reactions as well.

3. Advantages of microwave dielectric heating of microfluidic devices

Advantages of microwave dielectric heating include its preferential heating capability and non-contact delivery of energy. The first advantage stems from the fact that the microwave energy can be directly delivered to the microfluid sample with little or no absorption from the substrate material (glass, PDMS). The latter advantage facilitates not only the faster heating rates but also the faster cooling rates. These characteristics of microwave dielectric heating allow the application of heat-pulse approaching a delta function, because the heating stops at the moment the microwave power is turned off. This aspect of microwave heating has been used by (Fermer et al., 2003) and (Orrling et al., 2004) for high-speed polymerase chain reactions. The mechanisms of chemical reactions assisted by microwave heating have been found to conform close to theory yielding much more reliable end-products (Zhang et al., 2003; Whittaker et al., 2002; Gedye et al., 1998; Langa et al., 1997). The chemical reactions assisted by the microwave heating have also been performed at much lower temperatures compared to the conventional heating methods (Bengtson et al., 2002; Fermer et al., 2003). Localized microwave heating of fluids has also been demonstrated in the systems having silicon field-effect-transistors in the vicinity of microfluids (Elibol et al., 2008; Elibol et al., 2009).

Enhanced thermo-cycling rates and reduced reaction times compared to conventional techniques can be achieved because of the inertialess nature of microwave heating. Microwave-mediated thermocycling has been demonstrated for DNA amplification application (Kempitiya et al., 2009; Marchiarullo et al., 2007; Sklavounos et al., 2006). Heating can also be made spatially selective by confining the electromagnetic fields to specific regions of the microfluidic network. Further, the dielectric properties of the fluid can also be exploited to deliver heat using signal frequency as a control parameter in addition to the power.

Microwave heating is also very attractive than other alternatives for obtaining spatial temporal temperature gradients for a variety of on-chip applications, including investigation of thermophoresis (Duhr et al., 2006), control and measurement of enzymatic activity (Arata et al., 2005; Mao et al., 2002; Tanaka et al., 2000), investigation of the thermodynamics (Baaske et al., 2007; Mao et al., 2002), chemical separation (Buch et al., 2004; Huang et al., 2002; Ross et al., 2002; Zhang et al., 2007), and of the kinetics characterizing molecular associations (Braun et al., 2003; Dodge et al., 2004). Most techniques for generating on-chip temperature gradients integrate Joule heating elements to conduct heat into microchannels/microchambers (Arata et al., 2005; Buch et al., 2004, Selva et al., 2009). However, temporal control is limited by the heat capacity of the microfluidic device and thermal coupling of the device to the heating elements. It is easy to locally and rapidly generate temperature gradients within microchannels using the pattern of the microwave electric field intensity of a standing wave. The temperature distribution in the channel fluid is proportional to the time average of the square of the microwave electric field, which contains a sinusoidal component in the presence of a standing wave in the transmission line used to couple microwaves to the fluid in the microchannel. Using microwaves, a nonlinear sinusoidally shaped gradient along a channel of several millimeter length with a quasilinear temperature gradient can be achieved within a second (Shah et al., 2010) . The electric field distribution can also be controlled via the operating frequency and input power, which provides flexibility in changing the temperature profile for different specimens, reactions or applications.

4. Mechanism of dielectric heating of water-based fluids

4.1. Physical viewpoint

The water molecule has a permanent dipole (the central oxygen atom is electronegative compared to the hydrogen atoms, which are covalently bonded to the oxygen atom) and tends to align itself with an applied electric field. The resistance experienced by the water dipole molecule in aligning itself with the applied electric field is directly related to the intermolecular forces (hydrogen bonds formed by the oxygen atom of one water molecule with the hydrogen atoms of other water molecules) it encounters. Under an influence of sinusoidal applied electric field at a microwave frequency, the ensemble of water dipole molecules experience a rotational torque, in orienting themselves with the electric filed. The rotation caused by the applied field is constantly interrupted by collisions with neighbors. This process results in hydrogen bond breakage and the energy associated with the hydrogen bonds gets translated into the kinetic energy of the rotating dipoles. The higher the angular velocity of a rotating molecule, the higher the angular momentum, and consequently the higher is the kinetic energy. Thus, intermolecular collisions lead to friction, which causes dielectric heating. Dielectric heating is quantified by the imaginary part (ε'') of the dielectric constant. The value of ε'' (also called as the dielectric loss factor) depends on the frequency. As the frequency increases from the MHz range into the GHz range, the rotational torque exerted by the electric field increases; consequently, the angular velocity of the rotating dipoles increases, resulting in an increase in the value of ε''.

4.2. Mathematical viewpoint

The orientation of the molecular dipoles in response to the applied electric field results in the displacement of the charges, which generates displacement current according to the Maxwell-Ampere law. Dielectric heating is the result of interaction between the displacement current and the applied electric field. At low frequencies (MHz range) the molecular dipoles are able to follow the changes in polarity of the applied electric field (E). Thus, even though a displacement current (I) is generated, it is 90° out of phase with the applied electric field, resulting in a $E \times I = E \cdot I \cdot \cos(90^\circ) = 0$. Thus, no dielectric heating occurs at such frequencies. At frequencies ≥ 0.5 GHz, the molecular dipoles cannot keep pace with the rapidly changing polarity of the applied electric field, and hence the displacement current acquires a component, $I \cdot \sin\delta$, in phase with the applied electric field, where δ is the phase lag between the applied electric field and molecular dipole orientation. This results in a $E \times I = E \cdot I \cdot \cos(90^\circ - \delta) \neq 0$, and consequently dielectric heating. At very high frequencies (≥ 50 GHz), the field changes too quickly for the molecular dipoles to orient significantly, hence, the displacement current component in phase with the applied electric field vanishes. Consequently, the extent of dielectric heating decreases at high frequencies.

4.3. Dependence on ionicity of the fluid and temperature

The absorption depth of microwave power for liquid water in the 2 – 25 GHz region is few to several tens of centimeters (Jackson et al., 1975). Since the microchannels are typically 5 –

10 μm in depth, the electric field intensity is more or less constant throughout the microchannel.

The microwave power absorbed per unit volume in a dielectric material is given by:

$$P_v = \sigma E^2 \quad (1)$$

where

$$\sigma = 2\pi f \varepsilon_0 \varepsilon'' \quad (2)$$

is the dielectric conductivity of the material, f is the frequency in Hz, ε_0 is the permittivity of free space, ε'' is the imaginary part of the complex permittivity of the material (which depends on the frequency and temperature), and E is the electric field strength in V/m within the material. The ε'' is given by:

$$\varepsilon'' = [(\varepsilon_s - \varepsilon_\infty)\omega\tau] / \left(1 + \omega^2\tau^2\right) \quad (3)$$

where, τ is the relaxation time, ω = 2πf is the angular frequency, ε_s is the static field permittivity, and ε_∞ is the optical domain permittivity at frequencies much greater than the relaxation frequency (1/τ). The ε'' value first increases with increasing frequency reaching a peak value, before it starts decreasing with a further increase in frequency.

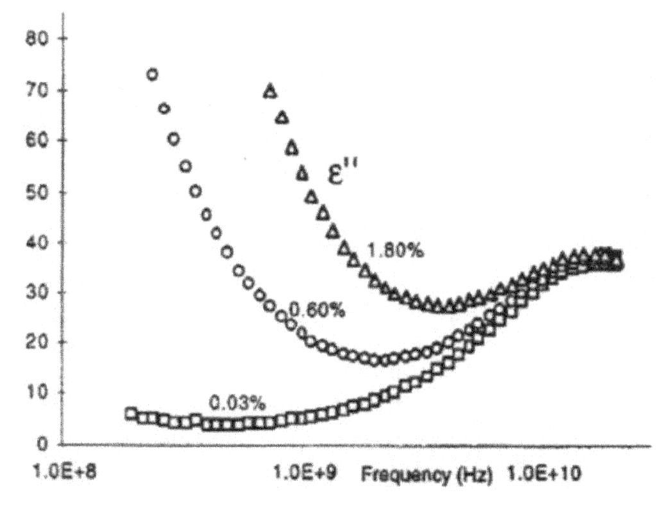

Figure 1. Experimental values for the permittivity and total loss factor as a function of frequency for aqueous KCl solutions of differing ionic concentrations (adapted from Gabriel et al., 1998).

The ions dissolved in water influence the rotational ability of the water molecules that are in close proximity to them. Under an external field, the mobile ions collide with the nearby water molecules transferring their kinetic energy to the water molecules, which is given out as heat. The ε" value, and consequently the dielectric conductivity, increase with increasing ionic strength at low frequencies, but interestingly, as shown in Figure 1, it becomes relatively independent of ion concentration of solution over a small region of microwave frequencies. Therefore, such solutions can be heated with microwaves of this frequency regime independently of their ionic strength. This characteristic of microwave dielectric heating is particularly advantageous when the salt concentration of the solution is not a known *priori* as would be a common case.

The water temperature also affects the microwave dielectric heating mechanism. With an increase in the water temperature, the strength and the extent of the hydrogen bonding network in water decreases, because more hydrogen bonds are already broken at a high temperature. This lowers the ε" value and consequently a decrease in dielectric heating. It means the water becomes a poorer absorber of microwave power with increasing temperature, shifting the ε" (or σ or P_v) versus frequency curve to higher frequencies. This can be an advantage when a steady temperature needs to be maintained.

4.4. Model of temperature gradient generation in microfluidic channel using microwaves

The power density in a dielectric material upon exposure to alternating electromagnetic field is given by (Woolley et al., 1996)

$$P = \omega \epsilon_0 \epsilon''(\omega) | E |^2 \tag{4}$$

where ω is the angular excitation frequency, ϵ_o is the vacuum permittivity, ϵ'' is the loss factor and E is the electric field strength in volts per meter within the material. For a wave traveling in the z-direction on a transmission line, the phasor representation of the total electric field is the sum of contributions from two separate components, the forward wave and the reflected wave, as described below:

$$E(z, t) = | E_+ | e^{-jkz} e^{j\omega t} + | E_- | e^{jkz} e^{j\theta p} e^{j\omega t} \tag{5}$$

where $jk = \alpha + j\beta$, E_+ is the amplitude of the forward wave, k is the complex propagation constant, E_- is the amplitude of the reflected wave, θ_p is the phase angle between the reflected and forward waves, α is the attenuation constant that describes the rate of decay of microwave power per unit length, z is the distance along the direction of propagation and β is the phase constant (change in phase per unit length) (Ramo et al., 1993). The time averaged power density (P) is proportional to $E(z, t) \cdot E(z, t)*$. Thus, we can compute the temperature profile of the fluid in the microchannel according to

$$T = a^2 e^{-2\alpha z} + b^2 e^{2\alpha z} + 2ab\cos(2 | \beta | z + \theta_p), \tag{6}$$

where T is the fluid temperature due to microwave heating. This simplified model describes several key features of microwave-induced temperature gradients. The dielectric properties of the transmission medium are non-homogeneous due to the presence of the microchannel resulting in impedance mismatch at the boundaries of the microchannel. The constructive and destructive interference caused by impedance mismatch between the forward and reflected waves at these boundaries generate a standing wave in the electric field and a corresponding stationary temperature field within the microchannel. The shape and magnitude of the temperature field depends on the microchannel geometry, the position of the microchannel relative to the transmission line, the frequency of operation and the input power. The rate of decay of the temperature field is governed by the transmission-line attenuation factor α, which is a function of the transmission-line geometry and the frequency-dependent loss factors of the transmission line materials. Hence, the higher the operating frequency of the microwave electric field, the lower the wavelength of the temperature field producing more peaks and valleys in the spatial temperature profile; and the higher the attenuation constant, the higher the average slope in the temperature field from the front to the back of the channel. Rest of the chapter focuses on macroscale and microscale types of microwave applicators for microfluidic heating applications.

5. Macroscale microwave applicators for microfluidic heating

The use of microwave heating has been demonstrated for a variety of applications including drug discovery, isolation of DNA, and heating of biological cells. Macro-scale microwave applicators were commonly utilized for the delivery of microwave power to sample contained in the plastic reaction tubes.

For microwave heating using a macroscale X-band rectangular waveguide, a slot is machined into one of the walls of the waveguide to allow the introduction of a microfluidic device, as shown in Figure 2. The position of the slot is chosen such that the microchip is placed perpendicular to the direction of the field propagation to maximize coupling of microwave power to fluid. The fluidic channel is micromachined into PMMA substrate using a milling machine. The channel enclosure is accomplished via thermal lamination technique with PP film. A traveling wave tube amplifier, capable of amplifying the input power up to 30 W over the 8 GHz to 18 GHz frequency range, coupled to a microwave signal generator is used to provide the desired microwave power; and a thermocouple inserted into the PMMA substrate is used to measure the fluid temperature. Further details of the system are given elsewhere (Shah, 2007b).

Thermo-cycling of de-ionized water between 60 °C and 95 °C is accomplished by this system, as shown in Figure 3. As mentioned earlier, DNA amplification by PCR relies on temperature cycling of the reaction mixture through three different temperatures between 50 °C and 95 °C. It means this system is suitable for DNA amplification by PCR. When a 20 W microwave power was applied the average heating rate of this system can be as high as 6 °C/s and the cooling rate 2.2 °C/s for a fluid volume of about 70 μL.

These values are better than what was accomplished (Figure 4) using a conventional metal block-based thermocycler whose heating and cooling rates are less than 1.7 °C/s. The results of Figure 3 confirm that microwave heating is a viable alternative for on-chip microfluidic systems, and that it can be used to obtain superior thermocycling rates compared to those obtained with conventional macroscale thermocyclers.

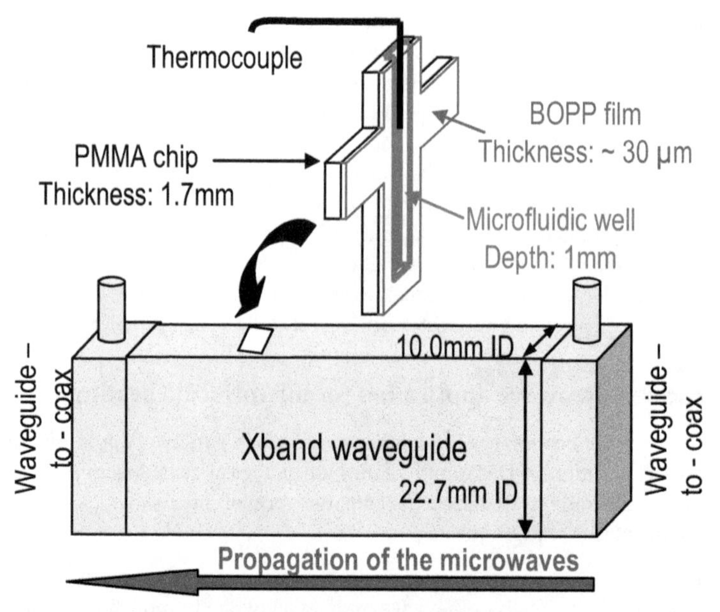

Figure 2. A schematic of an x-band rectangular waveguide heating system.

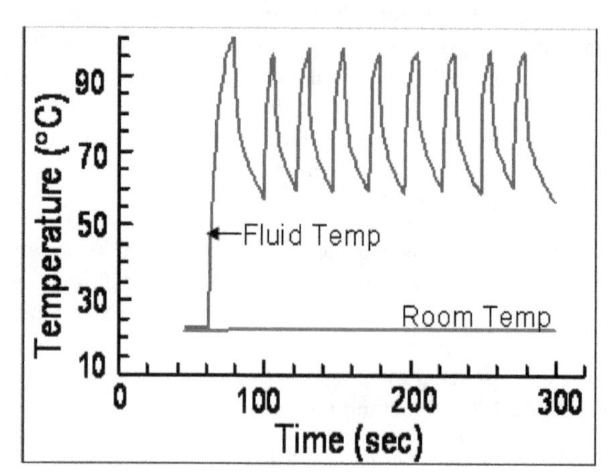

Figure 3. On-chip thermocycling of de-ionized water obtained using rectangular waveguide heating system.

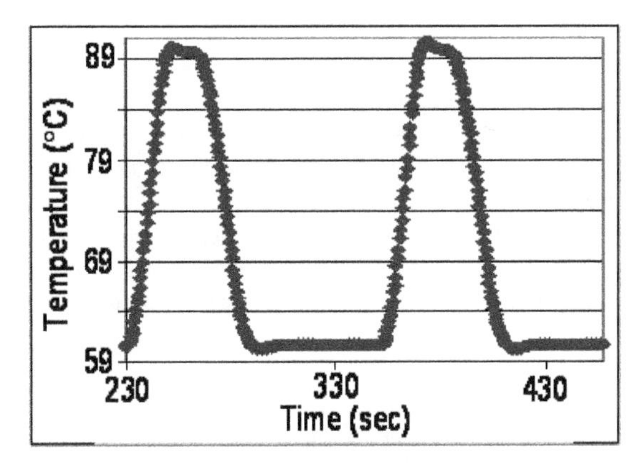

Figure 4. Thermocycling of de-ionized water using macroscale conventional thermocycler. The PCR tubes were used to hold 25 μL of fluid. The average heating rate for this system was 1.69 °C/s and the cooling rate was 1.36 °C/s.

6. Microscale microwave applicators for microfluidic heating

In recent years the interest in microwave heating lie in integrating miniaturized microwave heating elements, in the form of transmission lines, with microfluidic channels for on-chip heating applications. In one such case, microwave power can be delivered to an elastomeric microfluidic channel via an integrated thin-film coplanar waveguide (CPW) two-conductor transmission line. The rest of the chapter focuses on the work our group has done on this topic in the past 10 years.

6.1. Coplanar waveguide to deliver microwave power to the microfluidic channel

A schematic of a prototype coplanar waveguide (CPW) coupled with the microfluidic channel is shown in Figure 5. Typical dimensions are given in Figure 6. In this system a defined poly(dimethylsiloxane) (PDMS) microfluidic channel is aligned in the gap between the signal plane and one of the ground planes. The reason for choosing the CPW for delivering the microwave power to the microfluidic channel is that it can be easily formed on a substrate using a lift-off lithography process and its planar nature allows for easy integration with the microfluidic channels. The planar geometry also allows fabrication of an array of these devices on a common substrate with high yield, which later can be separated into individual devices.

The dimensions of the center conductor, the gap between the center conductor and the ground plane, the thickness and effective dielectric constant of the substrate determine the characteristic impedance (Z_O) and effective dielectric constant of the line. Since, the CPW is a two-conductor transmission line with different dielectrics above and below the device plane, it supports a quasi-TEM mode as well as surface wave modes (Riaziat et al., 1990). The surface wave modes lead to energy transfer from the guided wave into the substrate leading to the attenuation of the guided wave. The CPW design solution is optimized to avoid all potential problems associated with the surface waves in choosing the substrate thickness less than the dielectric substrate wavelength, λ_d (Riaziat et al., 1990). For the design features shown in Figure 6, the substrate thickness is one-sixth of the λ_d value. The conductor loss in CPW is inversely proportional to the skin depth and metallization thickness. This leads to increased conductor attenuation with increasing microwave frequency and decreasing conductor thickness, however, the effect of conductor thickness on the conductor attenuation is not as dramatic as that of frequency. For a glass substrate, the dielectric attenuation (Gupta et al., 1996) is very insignificant compared to the conductor attenuation and can be ignored.

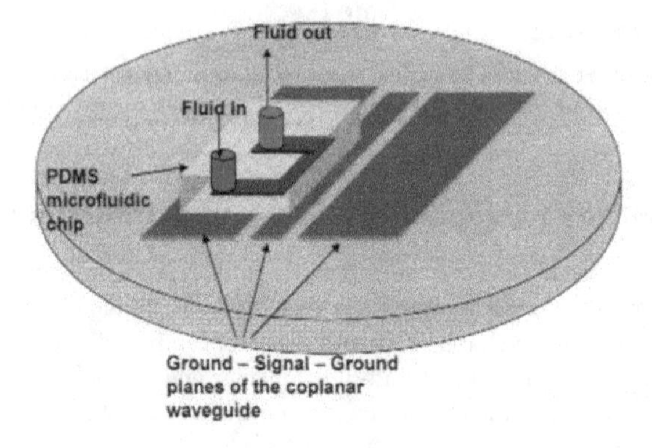

Figure 5. Schematic of coplanar waveguide (in orange) coupled with the microchannel structure (in pink)

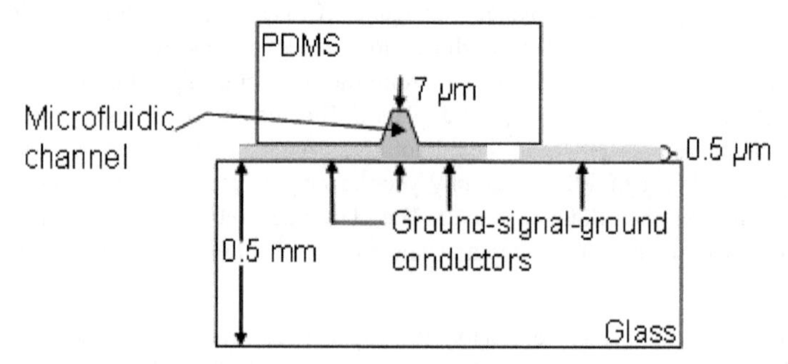

Figure 6. Schematic of a cross-section of a coplanar waveguide (CPW) transmission line integrated with a poly(dimethylsiloxane) (PDMS) microchannel for microwave dielectric heating of fluids. The CPW conductors are 1.5 cm long; the signal conductor is 140 µm wide and it is separated by a 25 µm gap on either side from 300 µm wide ground conductors. The microchannel consists of a trapezoidal cross section defined by a silicon template and it is 7 µm deep, 25 µm wide at the bottom, and 3.62 mm long.

The CPW conductors are formed of Cr/Au on the substrate using standard lift-off lithography metallization process. Thickness of the CPW conductors is increased by electroplating of Au. A trapezoidal cross-section PDMS microfluidic channel is defined by using a silicon template. The PDMS is poured into the silicon master and cured for 4 hours at 65 °C for this purpose. The PDMS structure is then detached from the silicon master and holes are punched into the PDMS structure for creating fluid inlet and outlet for the channel. The PDMS microchannel is then carefully aligned and attached by contact adhesion in the lateral direction over the gap between the signal-ground metal lines of the CPW. For maximum electric field coupling, the width of the PDMS microchannel is made equal to the CPW signal-ground metal spacing. According to computer simulations about 50% of the electric field is contained within 10 microns from the CPW surface. Hence, the depth of the microchannel should be designed accordingly to couple as much of the electric field as possible to the microfluid in the channel.

Microwave power is applied at the input port of the CPW using a signal generator. The frequency of the microwaves can be varied for optimum coupling of the microwave power to the microfluid. S-parameters of the system are measured for empty channel case and also for liquids of varying ionic concentrations. The fluid temperature is obtained by measuring the temperature dependent fluorescence intensity of a dilute fluorophore dye added to the fluid and comparing it to the calibrated fluorescence intensity at a known temperature. Rhodamyne B is used as the dye solution for this purpose.

6.2. Integrated microfluidic device for generating microwave-induced temperature gradients using a microstrip waveguide configuration

Figure 7 shows a cross-sectional schematic and a picture of the microwave-heated microfluidic device for generating temperature gradients. The devices are fabricated using an adhesive copper tape on cyclic olefin copolymer (COC) using photolithographic procedures. A CNC milling machine is used to cut the substrate material to precise chip dimensions and to carve out the microchannel and the fluidic access ports. For the device in Figure 7 the microchannel is 340 μm wide, 7 mm long and is machined all the way through a 300 μm thick COC substrate. The channel is positioned 1 cm away from the front of the device. The signal line is 370 μm wide and 5 cm long. The signal line and the ground plane are patterned on the copper tape after the tape is fixed on the top and bottom surfaces of the COC substrate. The tape has a 40 μm thick acrylic adhesive on a 35 μm thick copper foil. The acrylic adhesive serves as a cover plate for the microchannel and isolates the fluid from the copper electrodes. A slit, 100 μm wide and 2 cm long, is formed in the ground plane to allow optical detection of the fluid for temperature measurement. The dimensions of the microchannel, the transmission-line and the slit were determined from electromagnetic simulations performed using Sonnet Software. The slit is designed to run parallel to the microchannel, and it is positioned in the middle of the microchannel (Figure 7(*b*)) with a photolithographic process. Such positioning allows for spatially resolved, optical fluid-temperature measurements throughout the length of the microchannel and laterally in the middle of the channel. Two 2.4 mm end-launch connectors are mounted on the device

(Figure 7(*a*)) to convert from the microstrip geometry to the coaxial geometry of the test equipment.

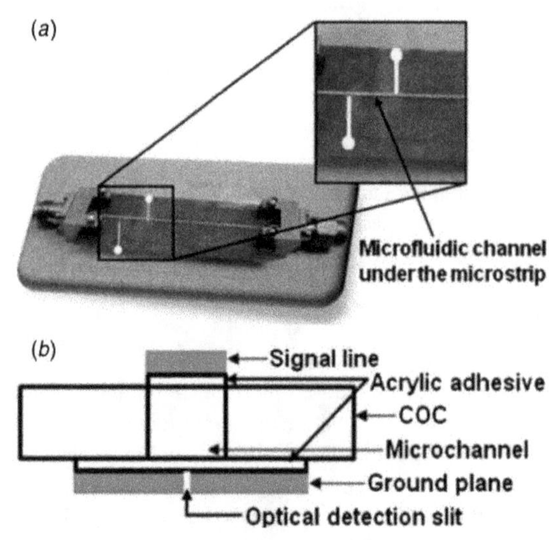

Figure 7. (a) A picture of the integrated microfluidic device for generating microwave-induced temperature gradients. (b) A cross-sectional view of the microwave heating device.

The method described here for fabricating conductors is easily transferable to other microfluidic substrates such as glass and PDMS. Furthermore, the one-step method for conductor fabrication obviates the need for electroplating, which is typically required following thin-film deposition to achieve sufficient conductor thickness. This method also provides easy bonding of the top and bottom cover plates to create enclosed channel structures, which has proven challenging for thermoplastic materials (Shah et al., 2006). In contrast to previously published reports (Shah et al., 2007a; Booth et al., 2006; Facer et al., 2001), the transmission line structure isolates the fluid from the metal conductors making these devices suitable for a variety of biochemical applications in which reagent contamination due to electrolysis or corrosion is undesirable.

6.3. Temperature measurement equations for Rhodamine B dye solution in the context of microfluidics

The temperature-sensitive nature of molecular fluorescence is suitable for measurement of temperature in small volume fluids like in microfluidic channels (Lou et al., 1999; Gallery et al., 1994; Sakakibara et al., 1993; Ali et al., 1990; Kubin et al., 1982). The temperature measurement is based on measuring fluorescence intensity ratios. The fluorescence intensity is typically measured at a known reference temperature, which is later used to normalize the intensity measured after heating the medium to an unknown temperature. The temperature is extracted by substituting the normalized intensity into a calibration curve. For lab-on-a-chip application, the use of dilute solutions of a temperature-sensitive fluorescent dye, particularly Rhodamine B (RhB), has become very popular for optical measurement of temperature distributions. The RhB is a water-soluble fluorescent compound with an absorption peak at 554 nm, an emission peak at 576 nm, and a temperature dependent fluorescence quantum yield (Ferguson et al., 1973). In addition to high temperature sensitivity, its other properties such as negligible pressure sensitivity and nominal pH-independent absorption (above a pH value of 6) are attractive for measuring temperatures in microfluidic channels with high spatial and temporal resolution (Ross et al., 2001). The RhB solution has been employed: to examine in-channel temperature and flow profiles at a T-shaped microchannel intersection during electrokinetic pumping (Erickson et al., 2003) and to characterize the temperature field resulting from resistive microheaters embedded in a poly(dimethylsiloxane) (PDMS) microchip (Fu et al., 2006). Even though the RhB has been primarily used for temperature measurement in aqueous environment, the absorbed RhB dye molecules in a PDMS thin film can be used for whole chip temperature measurement (Samy et al., 2008). The calibration equations used for computing temperature using fluorescent dyes relate the fluorescence intensity at an unknown temperature to the intensity at only one particular reference temperature. Such relations are not directly usable to applications requiring a different reference or initial temperature. The existing single-dye calibration equations have been generalized for extending their use to fluorescence intensity data normalized to reference temperatures other than those for which the original calibration equations were derived (Shah et al., 2009). Two methods have been described in detail: one is approximate, while the other, based on solution of a cubic equation, is an accurate mathematical treatment that does not incur errors beyond those already inherent in the calibration equations.

6.3.1. Generalization of RHODAMINE B temperature equations

Let $S(T)$ represent the signal received from a fluorescence detection system observing a small volume of fluorescent species at temperature T and let $I_{RT}(T) = S(T)/S(RT)$ represent the fluorescence intensity ratio normalized to the signal measured at nominal room temperature (RT).

	[RhB] (mmol/L)	[buffer] (mmol/L)	pH	A_0 (°C)	A_1 (°C)	A_2 (°C)	A_3 (°C)	T_0 (°C)	RT (°C)
Ross et al.	0.1	20	9.4	132	-250	220	-79	23	22
Fu et al.	0.05	25	8.5	149.15	-317.84	323.41	-131.84	22.88	23.5
Samy et al.	1.0,5.0	none	N/A	141.53	-250.25	228.02	-96.904	22.39	23

Table 1. Properties of Different RhB Solutions Used by Three Different Authors for Fluorescence-Based Temperature Measurements.

The fluorescence intensity acquired from an image captured at an elevated temperature, $S(T)$, is normalized by the intensity acquired from an image captured at nominal room temperature, $S(RT)$, to obtain $I_{RT}(T) = S(T)/S(RT)$. The temperature is then obtained by a least-squares adjustment of the constants in the equation

$$T = A_0 + A_1 I_{RT}(T) + A_2 I_{RT}^2(T) + A_3 I_{RT}^3(T) \quad (7)$$

to fit the measured $I_{RT}(T)$ for different values of T.

Table 1 includes the values of A_0 to A_3 that were reported by Fu et al., 2006; Ross et al., 2010; and Samy et al., 2008. The temperatures T_0 given by eqn. 7 when $I_{RT}(T) = 1.0$, as well as the RT used by each of the authors for normalization purposes are also shown in Table 1. These temperatures differ slightly from the normalized temperatures used by the different authors because their calibration equations were not constrained to produce RT when the intensity ratio was 1.0.

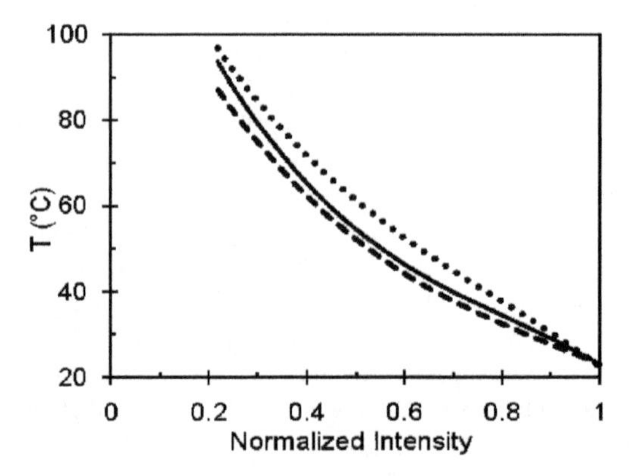

Figure 8. Comparison of the relative fluorescence-intensity versus temperature calibrations published by Fu et al., 2006 (—); Ross et al., 2010 (--); and Samy et al., 2008 (·····), for different rhodamine B chemistries

The calibration curves from all three authors are compared in Figure 8. The general trend of all three curves is similar. The difference between the curve of Samy et al., 2008 and the other two authors can be explained by the different physical medium and local environment used in the measurements.

6.3.2. Generalized calibration equation

Measurement of fluorescence intensity at a reference temperature in the vicinity of 23°C, for normalization purposes, is not always possible. For example, in applications requiring rapid temperature cycling of fluidic solutions the cycling temperatures of interest might be significantly different from 23 °C. In these situations, it is useful to calculate the fluorescence intensity ratio at a convenient reference temperature, T_1, in the temperature range of interest as

$$I_{T1} = I_{T1}(T) = S(T)/S(T1) \quad (8)$$

While it might appear plausible to use I_{T1} (T) in eqn. 7 instead of $I_{RT}(T)$ and then add $(T_1 - T_0)$ to the result to estimate the temperature, this process will add some additional error to the calculated temperatures obtained based on the calibration equations of Fu et al., 2006; Ross et al., 2010; and Samy et al., 2008 (Shah et al., 2009). The additional error introduced by this procedure is zero when $T_0 = T_1$ and larger than ±1 °C in certain temperature regions for some values of T_1 for all three calibration equations.

With a different approach, it was possible to eliminate all additional error except that inherent to the calibration equations themselves. For this approach, the normalized fluorescence intensity is generalized by rewriting it as

$$I_{RT}(T) = \frac{S(T)}{S(RT)} = \frac{S(T)}{S(T_1)}\frac{S(T_1)}{S(RT)} = I_{T_1}(T) I_{RT}(T_1) \quad (9)$$

where T_1 is any convenient known reference temperature. Therefore, if $I_{T1}(T)$ data have been measured where T_1 is not the reference temperature used in deriving the calibration equation, then $I_{RT}(T)$ can be calculated for use in eqn. 8 from eqn. 9, where $I_{RT}(T_1)$ can be obtained from the real solution(Terry et al., 1979) of the cubic equation

$$0 = A_3 I_{RT}{}^3(T_1) + A_2 I_{RT}{}^2(T_1) + A_1 I_{RT}(T_1) + A_0 - T_1 \quad (10)$$

with the values of A_n as listed in Table 1 and

$$I_{RT}(T_1) = \frac{-A_2}{3A_3} + (R + \sqrt{D})^{1/3} + (R - \sqrt{D})^{1/3}$$

$$D = Q^3 + R^2$$

$$Q = \frac{3A_1A_3 - A_2{}^2}{9A_3{}^2}$$

and

$$R = \frac{9A_1A_2A_3 - 27(A_0 - T_1)A_3{}^2 - 2A_2{}^3}{54A_3{}^3} \quad (11)$$

When $I_{RT}(T)$ is calculated from eqn. 9 with measured $I_{T1}(T)$ data and $I_{RT}(T_1)$ obtained from the solution to eqn. 10 that is given in eqn. 11, no error is introduced into the result beyond that already inherent in eqn. 7. A treatment similar to that described above can be applied to generalize calibration equations based on linear (Gallery et al., 1994) and second order polynomial (Erickson et al., 2003) fits to normalized $I(T)$ data to a convenient reference temperature.

6.4. Device characterization

Experiments were performed to evaluate the performance of the CPW devices for heating in the microchannel environment. The device is characterized for its frequency response in order to obtain absorption ratios with empty and fluid-filled microchannels. The fluid temperature is measured at various microwave frequencies. The results obtained from the first experiment are used to derive a power absorption model to find the distribution of the incident microwave power in different absorbing structures of the device.

The CPW device frequency response is characterized by scattering (S) parameter measurements. The S-parameters relate the forward and reflected traveling waves in a transmission medium and can be used to understand the power flow as a function of frequency. The S-parameters are used in conjunction with the conservation of energy to model absorption of microwave power and to predict the fluid temperature based on the absorbed power. The predicted temperature is then fitted to the measured temperature to determine heating efficiency. The experiments for the results presented below were performed with deionized (DI) H_2O and with fluids of two different salt concentrations: 0.9% NaCl solution and 3.5% NaCl solution.

6.4.1. S-parameter measurements

The amplitude of the reflection coefficients (S_{11}) and transmission coefficients (S_{21}) from 0.3 GHz to 40 GHz for the device with an empty channel as well as for the fluid-filled channels are shown in Figure 9(a) and Figure 9(b), respectively. In comparison with the empty channel device, S_{11} is reduced for the device with fluid-filled channel and it approaches that of an empty channel for frequencies above 10.5 GHz. The decrease in S_{11} of the fluid-filled devices below 10.5 GHz which indicate good impedance matching conditions are apparently fortuitous. The S_{11} is also found to be almost independent of the ionic concentration of the fluid. The localized peak and trough features observed in the S_{11} are likely interference effects due to reflections at the probe-CPW interfaces and the CPW/air-CPW/fluid interfaces (Facer et al., 2001). The S_{21}, as seen from Figure 9(b), decreases with

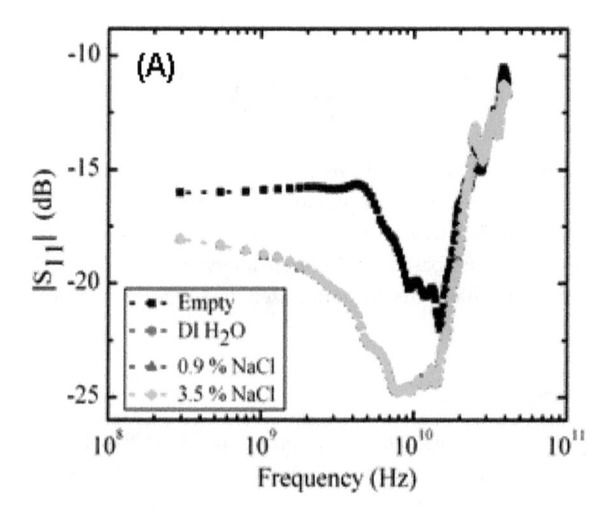

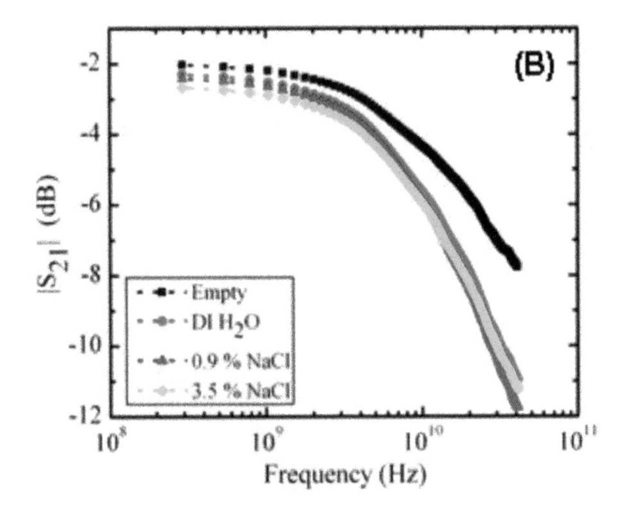

Figure 9. (A) Measured reflection coefficients ($|S_{11}|$)of the device. (B) Measured transmission coefficients ($|S_{21}|$)of the device. (■) Empty microchannel, (•) microchannel filled with deionized H_2O, (▲) microchannel filled with 0.9 % NaCl, (♦) microchannel filled with 3.5% NaCl.

increasing frequency for the devices with both the empty and fluid-filled channels. It is believed that the apparent low transmission coefficient of the empty-channel device is likely due to the smaller than optimum thickness of the CPW conductors (0.5 μm). The difference in S_{21} between the empty and fluid-filled channels becomes more pronounced at higher frequencies (> 5 GHz). The S_{21} of the fluid-filled devices is smaller compared to the empty-channel device due to the additional attenuation caused by the absorption of the microwave energy in the water.

Figure 10 shows percent absorption ratios (the fraction of the incident power absorbed by the device) as a function of frequency for the devices with empty and fluidfilled microchannels. The absorption ratio, A, is calculated from the S-parameter data shown in Figure 9 and using the equation: $A = 1 - R - T$ where R (the reflection coefficient) = $|S_{11}|$, T (the transmission coefficient) = $|S_{21}|$, and $S_{ij}\,(dB) = 10\,log_{10}|S_{ij}|$. It should be noted that the absorption ratio is dependent on the position of the microchannel over the length of the waveguide. In other words, S_{21} would not equal to S_{12} and the device would function as a non-reciprocal two-port network unless the microchannel is precisely centered over the length of the waveguide. The absorption ratio obtained for the device with empty channel (■) is due to ohmic dissipation in the thin-film CPW, which can be modeled by an attenuation constant, α_{cpw}, which corresponds to 2.86 dB/cm at 10 GHz. The attenuation constant of PDMS is assumed to be negligible because of the relatively low values of loss tangent (the ability of a material to convert stored electrical energy into heat) at microwave frequencies (Tiercelin et al., 2006). An increase in A observed with increasing frequency for the empty channel device (■) is expected from the dependence of skin depth on frequency. The absorption ratio for the fluid filled devices (•, ▲, and ♦) is greater for all frequencies measured in comparison with the empty channel device (■), due to microwave absorption in water and also ionic absorption in the ionic solutions. The data also exhibit a dependence of A on ionic concentration at lower frequencies as would be expected due to ionic conductivity, while the data at higher frequencies show that A is approximately independent of ionic concentration as would be expected due to dielectric conductivity. This trend is in agreement with theory (Wei et al., 1990).

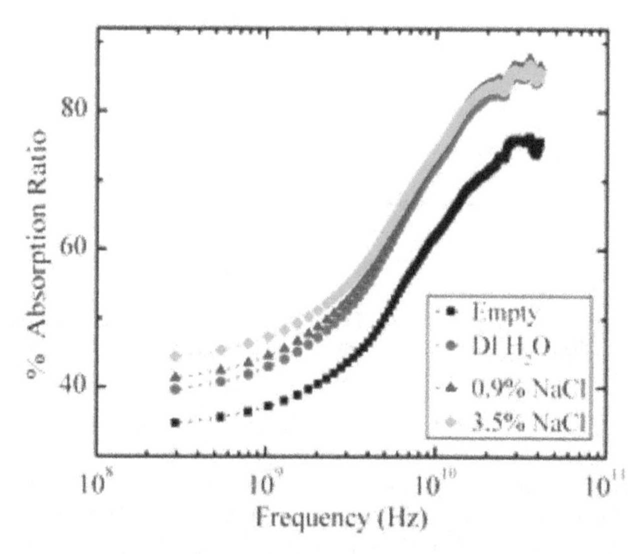

Figure 10. The percent absorption ratios (the fraction of the incident microwave power absorbed by the device) as a function of frequency. The absorption ratio, A, was calculated from the measured transmission and reflection coefficients using $A = 1 - R - T$. (■) Empty microchannel, (•) microchannel filled with deionized H_2O, (▲) microchannel filled with 0.9% NaCl, (♦) microchannel filled with 3.5% NaCl.

6.4.2. *Power absorption model*

It seems from Figure 10 that the simplest approximation to the fraction of the incident microwave power absorbed in the fluid is the difference between the power absorbed by the full-channel and empty-channel devices (for the case of water-filled device, $A_{H2O} - A_{empty}$ in Figure 10). However, various microwave power absorption models described in (Geist et al., 2007) show that this simple approximation greatly underestimates the actual fraction of incident microwave power absorbed in the fluid. For this reason, a simple model, alpha absorption model, is chosen to extract the fraction of the incident power absorbed in the fluid. This model is also used to differentiate microwave heating of the fluid from conductive heating due to ohmic heating of the CPW conductors. As shown in Figure 11, the model is constructed by assuming that the PDMS completely covers the CPW and by partitioning the CPW into three regions: a center region that interacts with the fluid in the

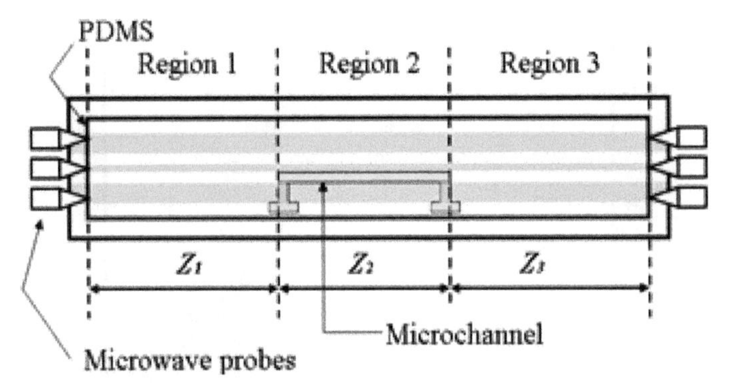

Figure 11. Top view of a CPW integrated with an elastometric microchannel consisting of three regions, center region with the microchannel and the two end regions without the microchannel.

microchannel and two end regions that have no microchannel over them. The lengths of each of the three regions are defined by the center region, Z_2, 0.36 cm long and two end regions, Z_1 and Z_3, each 0.57 cm long. The reflectance at the interface between the regions is assumed to be negligible, and the transmission coefficient, *T*, is modeled as follows:

$$T = (1 - R) \times e^{-2\alpha_{cpw} z_1} \times e^{-2\alpha_{cpw} z_2} \times e^{-2\alpha_w z_2} \times e^{-2\alpha_{cpw} z_3} \tag{12}$$

The derivation of this equation is based on exponential attenuation of microwave power in the direction of propagation where the rate of decay with distance is described by the attenuation constant, α. Because of the presence of water in the center region, it should be noted that the attenuation due to water, α_w, is added to the attenuation due to CPW, α_{cpw}. Eqn. 12 is then used to derive α_{cpw} and α_w as follows. First, α_{cpw} is derived for the empty channel device by setting α_w equal to zero into eqn. 12 and substituting the measured values of *T* and *R* for the empty channel device. Next, this value of α_{cpw} is substituted into eqn. 12 along with the measured values of T_f and R_f, which are *T* and *R* for the water filled device, respectively. The resulting equation is solved for α_w, which is found to be 3.68 dB/cm at 10 GHz. The absorption ratio for the central region of the water filled channel, A_2, is calculated as

$$A_2 = (1 - R_f) \times (e^{-2\alpha_{cpw} z_1}) \times (1 - e^{-2(\alpha_{cpw} + \alpha_w) z_2}) \tag{13}$$

Finally, the absorption ratios of the water, Aw, and CPW conductors, $Am2$, in the center region are calculated by eqn. 14 and eqn. 15, respectively.

$$A_w = \left[\frac{\alpha_w}{\alpha_{cpw} + \alpha_w} \right] A_2 \tag{14}$$

$$A_{m2} = A_2 - A_w \tag{15}$$

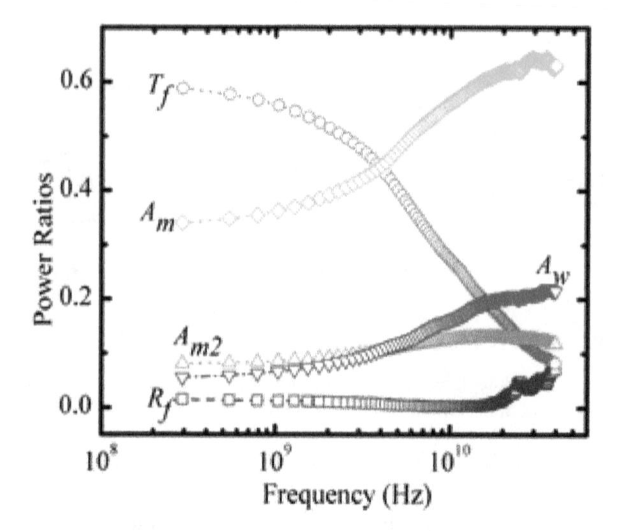

Figure 12. The distribution of the incident microwave power in different absorbing structures of deionized H_2O filled device as obtained from the alpha absorption model. T_f, R_f are the transmission and reflection coefficients of the water-filled device, respectively; A_m, A_{m2}, A_w are the absorption ratio of the CPW conductors, the absorption ratio of the CPW conductors in the region with the microchannel, and the absorption ratio of the water, respectively.

Eqns 12 through 15 are first order approximations. For impedance matched conditions, the equations show that A_w is dependent of position of the microchannel along the length of the CPW with a maximum occurring at the source end because the transmitted power attenuates exponentially along the transmission line. Further, A_w decreases exponentially along the length of the microchannel. Therefore, there is a design tradeoff between the microchannel length, its position relative to the source, and the frequency of operation to obtain a uniform temperature rise and efficient absorption of microwave power. Figure 12 shows the distribution of the incident microwave power in different absorbing structures of deionized H_2O filled device. It shows that A_w increases with frequency in agreement with theory because the power absorbed by water per unit volume is proportional to $f\varepsilon''$ as calculated from Franks (Franks, F., 1972) . It also shows that the absorption by the metal in the central region (A_{m2}) competes with the absorption by the water (A_w). At low frequencies (< 3.5 GHz), A_{m2} is slightly higher than A_w. The difference between the two becomes negligible as the frequency increases further, and A_w starts to dominate as the frequency exceeds approximately 8 GHz. This can be explained by the differences in the attenuation constants. For instance, α_w is 1.3 times as high as α_{cpw} at 10 GHz. It can be observed from Figure 12 that the fraction of the incident power absorbed by the CPW conductors, A_m, is noticeably high over the entire frequency range measured. This results in ohmic heating of the CPW conductors, which is also expected to contribute to the temperature rise of the fluid. However, based on the incident power absorbed by the CPW conductors in the center region, A_{m2}, a worst-case first order analytic calculation (Carslaw et al., 1959) of the contribution of the metal heating to water heating shows that the power dissipated in the CPW contributes less than 20 % of the total heating observed in the microchannel.

6.4.3. Temperature measurements

The points (■) in Figure 13 show the fluid (aqueous solution of Rhodamine B) temperature measured at various microwave frequencies. The applied power was kept constant at 10 mW. The temperature was measured ~ 250 ms after turning on the microwave power, which was approximated to be within 5% of quasi-thermal equilibrium. The error bars indicate the pooled standard deviation over all measurements for two instances at each frequency added with the estimated standard deviation (0.5 °C) of the room temperature (22.5 °C). The observed temperature rise was 0.88 °C/mW at 12 GHz and 0.95 °C/mW at 15 GHz.

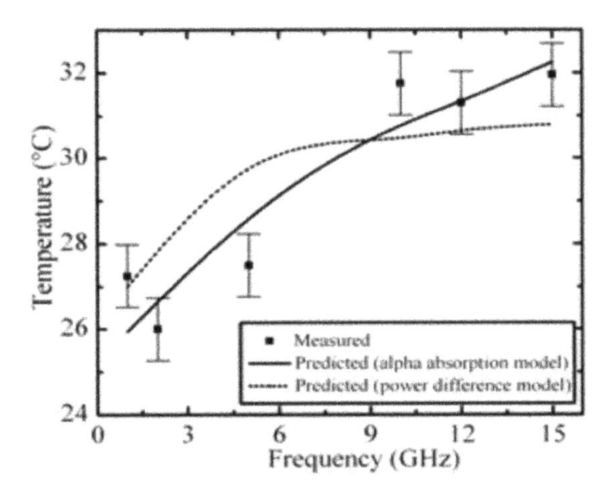

Figure 13. The measured temperature (■) of an aqueous solution of 0.2 mmol/L Rhodamine B in a 19 mmol/L carbonate buffer as a function of frequency. The solid line indicates predicted temperature calculated from the alpha absorption model, and the dashed line was calculated by employing the power difference model.

The temperature can also be calculated from the energy absorbed in the water during the heating period dt using equation 16

$$dT = \frac{P_v dt}{\rho C_p V} = \frac{K_e I A_w dt}{\rho C_p V} \tag{16}$$

Where ρ and Cp are the density and heat capacity of water, respectively, at appropriate temperature, I is the incident microwave power, A_w is the fraction of the incident power absorbed in the water as shown in Figure 12, and V is the volume of water in the microchannel. K_e is the channel-heating efficiency, which is defined as the fraction of energy absorbed in the water during the time dt that remains in the water. The rest of the energy absorbed in the water during the time dt is conducted into the substrate. The value of K_e cannot be easily obtained from the geometry and thermal properties of the channel and substrate due to the unknown contact thermal resistance (Kapitza resistance) between the water and the hydrophobic surface of the substrate (Barrat et al., 2003). Rather than attempt to calculate the value of K_e, it was adjusted in a least-squares fit of the predictions of eqn. 16 to the measured data points (■) in Figure 13. The solid line in Figure 13 indicates the predicted temperature calculated using A_w obtained from Figure 12 (alpha absorption model), and the dashed line indicates the predicted temperature calculated using Aw obtained from the power difference model ($A_{H2O} - A_{empty}$). It is clear from the results of fitting the predictions of the two different absorption models that the alpha absorption model provides a better fit to the measured data. Further, the heating efficiency obtained from the alpha absorption model indicates that only 5 % of the total heat (time integral of the absorbed power) was stored in the fluid while the rest was lost to the surroundings (PDMS and glass). Because the ratio of stored heat to lost heat increases with decreasing heating time, it is possible to confine most of the heat to the fluid and heat it to a higher temperature by increasing the microwave power and decreasing the heating period simultaneously.

6.4.4. *Nonlinear temperature gradients*

The choice of COC and Cu tape for the microfluidic cell (Figure 7) offers several advantages for producing ntegrated microfluidic devices for microwave heating. The high glass transition temperature of COC (T_g = 136 °C) as well as its chemical compatibility with acids, alcohols, bases and polar solvents make it suitable for photolithographic procedures. While the low thermal conductivity of COC (0.135 W m^{-1} K^{-1}) has a negative impact for contact heating approaches, it offers a significant advantage for direct volumetric-based heating strategies by minimizing undesired heat losses, so that a larger fraction of the incident power is contained in the fluid during heating. Additionally, the low dielectric constant (ε_r = 2.35) and the low loss factor (tan δ ~$1 \cdot 10^{-4}$ at 10 GHz) of COC make it suitable for high-frequency applications. The use of electro-deposited Cu compared to metal alloys as in the low-melt solder fill technique (Koh et al., 2003) e.g. a combination of indium-bismuth-tin alloy (Yang et al., 2002) for forming the transmissionline electrodes permits high-frequency operation of the devices due to the high electrical conductivity (σ_{Cu} ~5.51×10^5 S cm^{-1}, $\sigma_{In\text{-}alloy} = 0.19 \times 10^5$ S cm^{-1}) of Cu. On the other hand, if the conductor thickness (which is 35 μm in the device of Figure 7) exceed the thickness (3δ = 2 μm at 10 GHz for Cu, where δ is the skin depth) needed to sufficiently suppress ohmic losses due to the skin effect, then the greater-than-required thickness of Cu (33 μm in this case) acts as a thermal heat sink, limiting the maximum achievable temperature in the microchannel.

An electromagnetic simulation of a geometrical structure similar to that shown in Figure 7(*a*) was performed using Sonnet Software with nominal properties for copper, acrylic, COC and water. The design parameters were varied to optimize microwave power absorption in the fluid since absorption governs the maximum attainable temperature. A trade-off relation was found to exist between the power absorbed in the fluid and the ratio of the channel height to the total substrate thickness (the sum of cover plates and COC thickness). A smaller channel height to substrate thickness ratio reduced the absorbed power for a given fluid volume and incident microwave power. The coupling of the microwave power from the amplifier (less than, but approximately equal to 1W) to the transmission line and the microchannel was characterized theoretically and experimentally by measuring the reflection coefficient (S_{11}) and the transmission coefficient (S_{21}) and by calculating the absorption ratio ($A = 1 - |S_{11}|\ |S_{21}|$), which describes the fraction of the incident power absorbed by the device, where $S_{ij}\,(\text{dB}) = 10 \log_{10}|S_{ij}|$.

Figure 14 shows the simulated and measured absorption ratios for the empty-channel and water filled device. Close agreement is found in both the amplitude and shape for the full-channel device, but only in shape for the empty-channel device with the correlation coefficient of 0.98 for the empty-channel device and 0.96 for the water filled device. The amplitude deviation in the experimental absorption ratio for the empty-channel device can be attributed to imperfections in the as-fabricated conductor. However, this difference is much smaller for the water-filled device apparently because fluid absorption dominates the absorptive process. The S-parameter information was also used to select frequencies where large sinusoidally shaped temperature gradients were expected. Specifically, it was found that constructive interference exists between the traveling and reflected waves at a frequency corresponding to local maxima in the absorption ratio curve for the water-filled

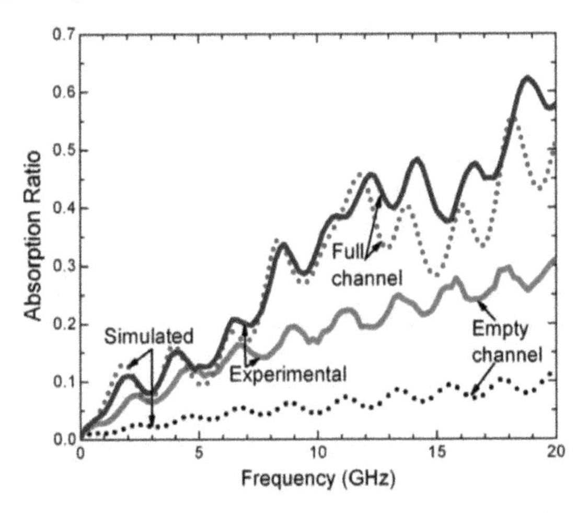

Figure 14. Comparison between simulated and measured absorption ratios, the fraction of the incident microwave power absorbed by the device, as a function of frequency. The simulated and measured responses are compared for the empty channel as well as the water-filled device. The curves are constructed from S-parameter data according to $A = 1 - |S_{11}| - |S_{21}|$, where S_{11} is the reflection coefficient and S_{21} is the transmission coefficient. The upper dotted line is simulated full channel response, the upper solid line is measured full channel response, the lower solid line is measured empty channel response, and the lower dotted line is the simulated empty channel response.

device. The absorbed power increases with increasing frequency, and the peaks in the absorbed power exist at a variety of frequencies (Figure 14). However, the amplitude of the absorbed power at the peaks for frequencies lower than 12 GHz is significantly smaller than that at 12 GHz and above. Here, the results of temperature measurements at the lowest (12 GHz) and highest (19 GHz) frequencies that gave relatively large peaks in the absorption ratio data are shown.

Figure 15 shows the experimental temperature profile for the excitation frequency of 19 GHz. The curve in Figure 15 was constructed by using the calibration curve of Rhodamine B dye to convert the fluorescence intensity into temperature. A nonlinearly modulated profile extending along the length of the microchannel was observed. The temperatures measured at different positions was compared with the model of temperature gradient generation by performing nonlinear least-squares fitting of equation 6 to the measured data points using Origin software (solid line in Figure 15) and a good agreement ($R^2 = 0.98$) was found. The parameter estimates and the associated standard errors are listed in Table 2.

	12 GHz		19 GHz	
	Value	SD	Value	SD
a (V mm^{-1})	6.986	0.051	6.31	0.024
b (V mm^{-1})	0.501	0.041	-0.281	0.017
β (deg mm^{-1})	29.11	0	46.35	0.974
θ_p (deg)	264.8	3.25	154.1	7.071
α (mm^{-1})	0.015	0.002	0.01	0.001
Chi-square	1.193	-	0.257	-
R^2	0.975	-	0.981	-

Table 2. Results from nonlinear least-squares fitting of the temperature gradient model (equation 6) to the measured data points shown in Figures 15 and 16 for one device. A standard error of zero indicates that this value was fixed during the fit.

	12 GHz		19 GHz	
	Mean	Average SD	Mean	Average SD
a (V mm^{-1})	6.912	0.049	6.427	0.035
b (V mm^{-1})	0.502	0.04	0.092	0.025
β (deg mm^{-1})	29.106	0.04	46.123	1.833
θp (deg)	259.4	3.229	193.6	14.22
α (mm^{-1})	0.011	0.002	0.012	0.002
Chi-square	1.142	-	0.712	-
R^2	0.971	-	0.923	-

Table 3. The average and standard deviations of the fitting parameters extracted from nonlinear least-squares fitting of measured temperature gradient data to the theoretical model shown in equation (6). A standard deviation of zero indicates that this value was fixed during the fit as described in the text.

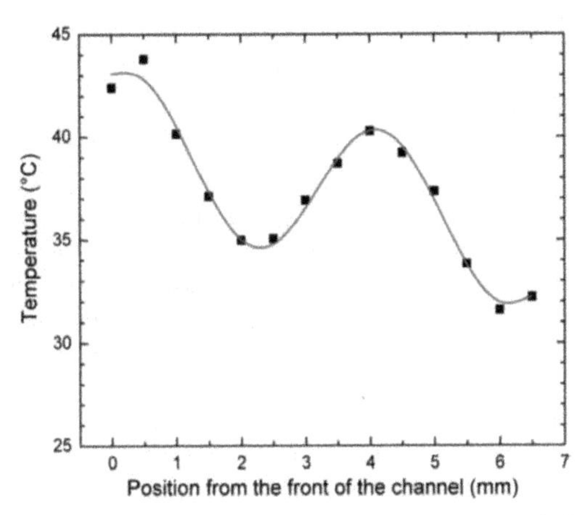

Figure 15. The measured temperature versus distance along the microchannel of an aqueous solution of 0.2 mmol L^{-1} Rhodamine B in a 19 mmol L^{-1} carbonate buffer at the microwave excitation frequency of 19 GHz. The solid line represents a theoretical temperature fit to the measured data points shown by squares. The measurement frequency was selected based on a local maximum in *A* for the full channel device. At 19 GHz, $S_{11} = 0.040$, $S_{21} = 0.343$ and $A = 0.617$ for the full channel device.

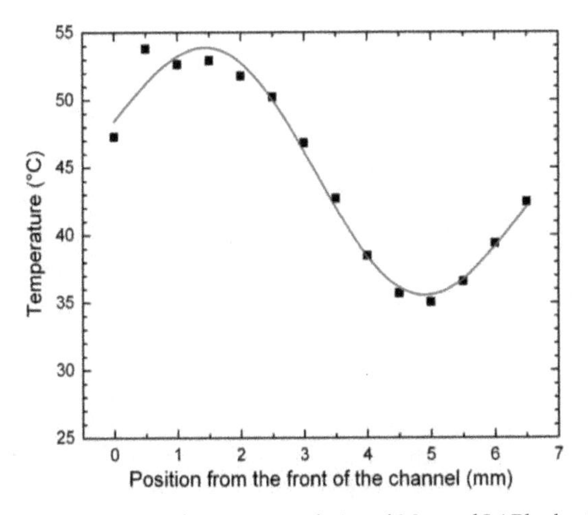

Figure 16. The measured temperature of an aqueous solution of 0.2 mmol L^{-1} Rhodamine B in a 19 mmol L^{-1} carbonate buffer as a function of position along the microchannel at the microwave excitation frequency of 12 GHz. The solid line represents a theoretical temperature fit to the data points shown by squares. The measurement frequency was selected based on a local minimum in S_{11} (not shown here) and a local maximum in *A* for the full channel device. At 12 GHz, $S_{11} = 0.063$, $S_{21} = 0.490$ and $A = 0.447$ for the full channel device.

For a given geometrical structure, the nonlinear temperature profile (Figure 15) can be altered by changing the frequency of the microwave signal. This is demonstrated in Figure 16, which shows the spatial temperature profile obtained for 12 GHz excitation frequency. Here, a nonlinear profile representing a sinusoidal wave extending along the length of the microchannel was observed. The data also resulted in a quasilinear temperature gradient with a slope of 7.3 °C mm^{-1} along a 2 mm distance. Linear temperature gradients with comparable slopes have been used for DNA mutation detection (Bienvenue et al., 2006), phase transition measurements in phospholipid membranes (Saiki et al., 1986), single-nucleotide polymorphism (SNP) analysis (Baker., 1995) and continuous-flow thermal gradient PCR (Liu et al., 2007). As before, nonlinear least-squares fitting of the theoretical model (solid line in Figure 16) to the experimental data shows good agreement (R^2 = 0.98). Due to limitations of the temperature gradient model, the fit to the 12 GHz data was not able to estimate the attenuation constant, α, with reasonable uncertainty because of multicollinearity (all of the coefficients of the covariance matrix were >0.8). Therefore, the fit to 12 GHz data by fixing β (Table 2) has been performed. The value of β was obtained from the fit to 19 GHz data by assuming that dispersion was negligible and scaling appropriately for the ratio of the two frequencies. One noteworthy application of a rapid, nonlinear temperature (electric field) gradient is electric field gradient focusing (EFGF) of charged molecules. It has been suggested theoretically that the peak capacity and resolution of EFGF and other related methods, such as TGF (Lagally et al., 2004), could be increased by using a nonlinear field (temperature) gradient provided that the first portion of the gradient is steep, the following section is shallow and that the sample components can be moved from the first portion to the second after focusing in the first portion (Chaudhari et al., 1998; Yoon et al., 2002). To demonstrate the efficacy of the technique for generating temperature gradients, measurements on several different devices were performed. The general shape of the temperature gradient curve was found to reproduce well for all of the measurements. The statistical deviations shown in Table 3 come primarily from variations in the geometrical dimensions of the device introduced during the fabrication process. Even though Rhodamine B dye was used successfully to demonstrate the presence of temperature gradients, accurate generation of the temperature gradient profile warrants improvements to the temperature detection method. In fact, a number of researchers have reported the specific absorption of Rhodamine B in PDMS substrates (Pal et al., 2004; Kopp et al., 1998). For the present device structure, the Rhodamine B was found to absorb into acrylic adhesive resulting in temperatures that were representative of channel surface rather than that of the bulk fluid. Additionally, after repeated use the absorption resulted in non-uniform fluorescence intensity across the length of the microchannel indicating that the absorption was non-uniform in space, which limits the reusability of the devices. Hence, these results demonstrate that while the method for generating spatial temperature gradients is robust as shown from the repeated measurements and low statistical error (Table 3), the frequent use of devices results in the channel surface becoming saturated with Rhodamine B dye limiting their overall use.

It should be possible to prevent the absorption of Rhodamine B dye on the microchannel surface by modifying those surfaces with appropriate surface treatments as is typically done for analyte separations in microfluidic devices (Schneegass et al., 2001). Further, a two-step process could be utilized to eliminate potential interactions of Rhodamine B dye with chemicals of interest. As the first step, a set of devices would be used with a dye solution to calibrate the temperature difference versus the microwave power characteristic of the device, and an identical device would be later used without the dye solution for performing biological or chemical studies. Alternatively, an electronic temperature sensor such as a thermocouple or resistance thermometer could be integrated into the device at a convenient reference location along the channel for optical calibration. Figure 17 plots the temperature as a function of time at one location along the microfluidic channel for a 1 s duration pulse of approximately 1 W of microwave power applied to the device. Substantially, more power (about 1 W from the amplifier) was required to raise the temperature of the fluid to 46 °C in 1 s than would be required to hold it at this temperature for an additional second as might be required in practical applications. It would be much easier to add feedback control of the microwave power if the temperature at a reference location was measured electronically rather than optically even if Rhodamine B or some other fluorescent dye was compatible with the other chemicals in the microchannel.

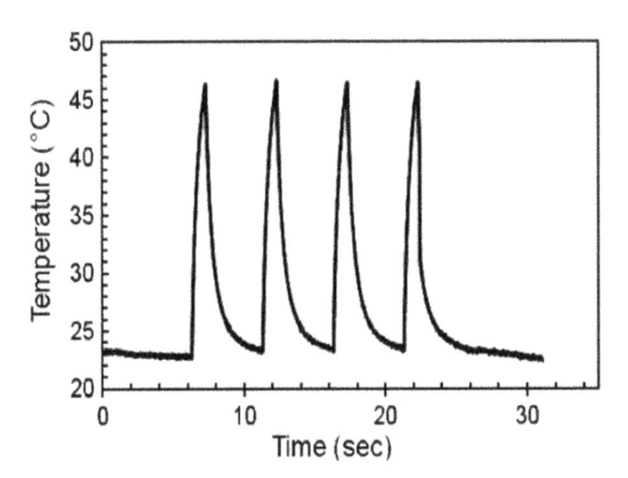

Figure 17. Transient temperature response of the integrated microfluidic device. A temperature of 46 °C was obtained at some locations in the microchannel with 1W of microwave power at the output of power amplifier for 1s. The fact that the temperature did not reach a steady state value in this time shows that considerably less power would be required to hold the temperature constant for a second. The addition of feedback control and a higher power amplifier would facilitate higher temperatures, a faster rate of increase in temperature, as well as the capability to hold the fluid temperature constant for a short period of time without raising the device temperature significantly.

Finally, the growing concern that exposure to microwaves can be harmful to living cells may limit the ability to operate highly integrated lab-on-a-chip devices containing living microorganisms in conventional microwave ovens. On the other hand, the microwave field decreases rather rapidly away from a properly designed microscale microwave generator, potentially allowing live organisms and microwave transducers to co-exist on a lab-on-a-chip device. The approach explained to establish temperature gradients appears to be especially well suited for thermal gradient focusing methods for analyte separations of cell metabolites in lab-on-a-chip devices. Other potential applications of integrated microwave heaters include cell lysis and PCR (Liu et al., 2002), as mentioned previously. The localized nature of on-chip microwave heating means that separate microwave heaters optimized for these different tasks could also co-exist on a single lab-on-a-chip device. Therefore, it is believed that the technique outlined here will facilitate the application of microfluidics to other biological and chemical applications requiring spatial temperature gradients as well as to temperature gradient generation.

7. Conclusions

The ability to rapidly and selectively control temperature within microchannel environment is crucial for many fluidic applications including high-efficiency PCR and temperature induced cell lysing. In this chapter, localized microwave dielectric heating of fluids at micrometer scale geometry using integrated planar microwave transmission line as a power source has been presented. The measured temperature increases with increasing frequency is in agreement with theory. The device offers several advantages. It is simple, easy to use and fabricate. The planar structure of the transmission line used as a power source lends itself to easy integration with the microchannel and allows for optical integration of the microchannel with widely used fluorescence microscopy techniques. The performance of the CPW for heating could be substantially improved by: applying high power pulses over shorter heating times, using a lower thermal diffusivity substrate than glass, and increasing the gap width and thickness of the CPW conductors.

A microwave power absorption model to understand power distribution through the device and to differentiate microwave heating of the fluid from conductive heating of the fluid because of microwave power absorption in thin-film CPW conductors has been presented. Based on the fitting of the experimental data using the power absorption model, it can be stated that the temperature rise of fluid is predominantly due to the absorbed microwave power. It is believed that this model can be useful for evaluating the performance of various complex and lossy transmission line configurations, such as CPW or microstrip lines, for heating fluid in the microchannel environment. Tt is believed that the microwave dielectric heating approach will be of particular use in rapid thermocycling applications and will lead to new applications exploiting heating in microfluidic environment. Such heaters are potentially very useful for single use, disposable, and integrated microfluidic systems.

The integrated microwave microstrip transmission line is also very attractive to generate temperature gradients rapidly and selectively in a microfluidic device. The shape of the temperature profile can be adjusted by varying the microwave excitation frequency and the amplitude of the profile can be adjusted by varying the microwave power. With this technique, the temperature gradients can be established locally and selectively by positioning the transmission line in the region of interest over the fluidic network. This device offers several advantages. Because the heating elements are integrated with the microchannel, this device offers a portable platform for generating spatial temperature gradients. The device is simple, easy to integrate and use with microfluidic channels, and allows for high frequency operation. The heating elements are easy to fabricate and the fabrication method is transferable to other microfluidic substrates. Reducing the thickness of the copper electrodes would provide larger temperature changes within the microchannel. This approach can be scaled for high throughput studies by fabricating multiple transmission lines in parallel. This approach to establish temperature gradients would be especially well suited for field gradient focusing methods for analyte separations. It will also facilitate the application of microfluidics to a multitude of other biological and chemical applications requiring spatial temperature gradients.

Author details

Mulpuri V.Rao
Department of Electrical and Computer Engineering, George Mason University,Fairfax, VA, USA

Jayna J. Shah
National Institute of Standards and Technology, Semiconductor Electronics Division, Gaithersburg, MD, USA
Presently at: Raytheon Company, Dallas, TX, USA

Jon Geist and Michael Gaitan
National Institute of Standards and Technology, Semiconductor Electronics Division, Gaithersburg, MD, USA

Acknowledgement

Authors acknowledge the contributions of Siddarth Sundaresan in the early part of this work on microwave dielectric heating of microfluidic devices.

8. References

Adnadjevic, B., JOvanovic, J. The Effect of Microwave heating on the Isothermal Kinetics of Chemicals reaction and Physicochemical Processes. In: Advances in Induction and Microwave heating of Mineral and Organic Materials. Intech; 2011

Arata, H. F., Rondelez, Y., Noji, H., Fujita, H.(2005). Temperature alternation by an on-chip microheater to reveal enzymatic activity of beta-galactosidase at high temperatures. Analytical Chemistry, Vol. 77, pp. 4810–4814

Ali, M. A., Moghaddasi, J., Ahmed, S. A. (1990). Examination of temperature effects on the lasing characteristics of rhodamine cw dye lasers. Applied Optics, vol. 29, pp. 3945-3949.

Baaske, P., Duhr, S., Braun, D. (2007). Melting curve analysis in a snapshot. Applied Physics Letters, Vol. 91, pp. 133901 1 – 3

Baker, D.R., Capillary Electrophoresis, John Wiley & Sons, Inc; 1995
Barrat, J. L.; Chiaruttini, F. (2003). Kapitza resistance at the liquid-solid interface. Molecular Physics, Vol. 101, pp. 1605-1610.
Bengtson, A., Hallberg, A., Larhed, M. (2002). Fast Synthesis of Aryl triflates with controlled microwave heating. Organic Letters, Vol. 4, No. 7, pp. 1231-1233
Bienvenue, J.M., Duncalf, N., Marchiarullo, D., Ferrance, J.P., Landers, J.P.(2006). Microchip-Based Cell Lysis and DNA Extraction from Sperm Cells for Application to Forensic Analysis . Journal of Forensic Science, Vol. 51, pp. 266-273.
Brodie, G., Microwave Heating in Moist Materials. In: Advances in Induction and Microwave Heating of Mineral and Organic Materials. InTech; 2011, pp. 553
Booth, J. C., Mateu, J., Janezic, M., Baker-Jarvis, J., Beall, J. A. (2006). Broadband permittivity measurements of liquid and biological samples using microfluidic channels. IEEE MTT-S International, Microwave Symposium. pp. 1750-1753
Braun, D., Libchaber, A. (2003). Lock-in by molecular multiplication. Applied Physics Letters, Vol. 83, pp. 5554–5556
Buch, J. S., Kimball, C., Rosenberger, F., Highsmith, W. E., DeVoe, D. L., Lee, C. S. (2004). DNA mutation detection in a polymer microfluidic network using temperature gradient gel electrophoresis. Analytical Chemistry, Vol. 76, pp. 874–881.
Carslaw, H. S.; Jaeger, J. C. (1959). Conduction of heat in solids. 2nd edition. Oxford University Press, Oxford.
Chaudhari, A. M., Woudenberg, T. M., Albin, M., Goodson, K. E., (1998). Transient liquid crystal thermometry of microfabricated PCR vessel arrays. Journal of Microelectromechanical Systems, vol. 7, pp. 345-355
Dodge, A., Turcatti, G., Lawrence, I., de Rooij, N.F., Verpoorte, E. (2004). A microfluidic platform using molecular beacon-based temperature calibration for thermal dehybridization of surface-bound DNA. Analytical Chemistry, Vol. 76, pp. 1778–1787
Duhr, S., Braun, D., (2006). Why molecules move along a temperature gradient. Proceedings of National Academy of Sciences, USA, Vol. 103, pp. 19678–19682
Elibol, O.H., Reddy, B., Bashir, R. (2008). Localized heating and thermal characterization of high electrical resistivity silicon-on-insulator sensors using nematic liquid crystals. Applied Physics Letters, Vol. 93 131908
Elibol, O.H., Reddy, B., Bashir, R. (2009). Localized heating on silicon field effect transistors: device fabrication and temperature measurements in fluid. Lab Chip, Vol. 9. pp 2789-2795
Erickson, D.; Sinton, D.; Li, D. Q. (2003). Joule heating and heat transfer in poly(dimethylsiloxane) microfludic systems. Lab on a Chip, Vol. 3, pp. 141–149
Facer, G. R., Notterman, D. A., Sohn, L. L. (2001). Dielectric spectroscopy for bioanalysis: From 40 Hz to 26.5 GHz in a microfabricated wave guide. Applied Physics Letter, Vol. 78 pp. 996– 998

Ferguson, J.; Mau, A. W. H. (1973). Spontaneous and stimulated emission from dyes. Spectroscopy of the neutral molecules of acridine orange, proflavine and rhodamine B. Australian Journal of Chemistry. Vol. 26, pp. 1617–1624

Fermer, C., Nilsson, P., Larhead, M. (2003). Microwave-assisted high speed PCR. European Journal of Pharmaceutical Sciences, Vol. 18, pp. 129-132

Franks, F. Water: a comprehensive treatise: The physics and physical chemistry of water. Plenum Press: 1972

Fu, R.; Xu, B.; Li, D. (2006). Study of temperature field in microchannels of a PDMS chip with embedded local heater using temperature-dependent fluorescent dye. International Journal of Thermal Sciences. Vol. 45, pp. 841–847

Gallery, J.; Gouterman, M.; Callis, J.; Khalil, G.; McLachlan, B.; Bell, J. (1994). Luminescent thermometry for aerodynamic measurements. Review of Scientific Instruments. Vol. 65, pp. 712–720

Gabriel, C., Gabriel, S., Grant, E. H., Halstead, B. S. J., Mingos, D. M. P. (1998). Dielectric parameters relevant to microwave dielectric heating. Chemical Society Review, Vol. 27, pp. 213-224

Gedye, R.N., Wei, J.B. (1998). Rate enhancement of organic reactions by microwave at atmospheric pressure. Canadian Journal of Chemistry, Vol. 76, pp. 525-532.

Geist, J.; Shah, J.; Rao, M. V.; Gaitan, M. (2007). Microwave power absorption in Low-Reflectance, Complex, Lossy Transmission Lines. Journal of Research of the National Institute of Standards and technology, Vol. 112, pp. 177-189

Gupta, K.C. Microstrips Lines and Slotlines, Artech House; 1996

Huang, T. M., Pawliszyn, J. (2002). Microfabrication of a tapered channel for isoelectric focusing with thermally generated pH gradient. Electrophoresis, Vol. 23, pp. 3504–3510

Jackson, J.D. Classical Electrodynamics, John Wiley & Sons; 1975

Kempitiya, A., Borca-Tasciuc, D.A., Mohamed, H.S., Hella, M.M. (2009). Localized microwave heating in microwells for parallel DNA amplification applications Applied Physics Letters, Vol. 94 064106

Koh, C. G., Tan, W., Zhao, M, Q., Ricco, A. J., Fan, Z. H. (2003). Integrating polymerase chain reaction, valving and electrophoresis in a plastic device for bacterial detection. Analytical Chemistry, vol. 75, pp. 4591-4598

Kopp, M. U., De Mello, A. J., Manz, A. (1998). Chemical Amplification: Continuous-Flow PCR on a Chip. Science, Vol. 280 pp. 1046 –1048.

Kubin, R. F., Fletcher, A.N., (1982). Fluorescence quantum yields of some rhodamine dyes. Journal of Luminescence, Vol. 27, pp. 455-462

Lagally, E.T., Mathies, R.A. (2004). Integrated genetic analysis Microsystems. Journal of Physics D: Applied Physics, Vol. 37, pp. R245-R261.

Langa, F.; de la Cruz, P.; de la Hoz, A.; Diaz-Ortiz, A.; Diez-Barra, E. (1997). Microwave irradiation: more than just a method for accelerating reactions. Contemporary Organic Synthesis, Vol. 4, pp. 373-386.

Liu, J., Enzelberger, M., Quake, S.R. (2002) A Nanoliter Rotary Device for Polymerase Chain Reaction. *Electrophoresis, Vol.* 23, pp. 1531-1536

Liu, P., Seo, T.S., Beyor, N., Shin, K.J., Scherer, J.R., and Mathies, R.A. (2007). Integrated portable polymerase chain reaction-capillary electrophoresis microsystem for rapid forensic short tandem repeat typing. Analytical Chemistry, Vol. 79, pp. 1881–1889.

Lou, J. F., T. M. Finegan., P. Mohsen., T. A. Hatton., P. E. Laibinis. (1999). Fluorescence-based thermometry: Principles and applications. Rev Analytical Chemestry, Vol. 18, pp. 235-284

Manz, A., Graber, N., Widmer, H.M. (2010). Miniaturized total chemical analysis systems: A Novel concept of chemical sensing. Senosors and Actuators B-Chemical, Vol. 1 pp. 244-248Mao, H. B., Yang, T. L., Cremer, P. S. (2002). A microfluidic device with a linear temperature gradient for parallel and combinatorial measurements. Journal of American Chemical Society, Vol. 124, pp. 4432–4435

Mao, H. B., Holden, M. A., You, M., Cremer, P. S. (2002). Reusable platforms for high-throughput on-chip temperature gradient assays. Analytical Chemistry, Vol. 74, pp. 5071–5075

Marchiarullo, D. J., Sklavounos, A., Barker, N.S., Landers, J. P. (2007). Microwave-mediated microchip thermocycling: pathway to an inexpensive, handheld real-time PCR instrument. 11th Int. Conf. on Miniaturized Systems for Chemistry and Life Sciences

Orrling, K., Nilsson, P., Gullberg, M., Larhed, M. (2004). An efficient method to perform milliliter- scale PCR utilizing highly controlled microwave themocycling. Chemical Communication, Vol. 7, pp. 790-791

Pal, R., Yang, M., Johnson, B.N., Burke, D.T., Burns, M.A. (2004). Phase change microvalve for integrated devices. Analytical Chemistry, Vol. 76, pp. 3740-3748

Ramo, S., Whinnery, J. R. Duzer, V. T. (1993). Fields and Waves in Communication Electronics 3rd edition (Hoboken, NJ: Wiley)

Reyes, D.R., Iossifidis, D., Auroux, P.A., Manz, A. (2002). Micro total analysis systems.1. Introduction, theory and technology. Analytical Chemistry, Vol. 74, pp. 2623-2636

Riaziat, M., Center, V. R., Alto, P. (1990). Propogation modes and dispersion characteristics of coplanar waveguides. IEEE transactions on Microwave Theory and Techniques, Vol. 38, pp. 245-251.

Ross, D.; Gaitan, M.; Locascio, L. E. (2001). Temperature measurement in microfluidic systems using a temperature-dependent fluorescent dye. Analytical Chemistry, Vol. 73, pp. 4117–4123

Ross, D., Locascio, L. E. (2002). Microfluidic temperature gradient focusing. Analytical Chemistry, Vol. 74, pp. 2556–2264

Saiki, R.K., Bugawan, T.L., Horn, G.T., Mullis, K.B., Erlich, H.A. (1986). Analysis of enzymatically amplified beta-globin and HLA-DQ alpha DNA with allele-specific oligonucleotide probes. Nature, Vol. 324, pp. 163–166.

Sakakibara, J., Hishida, K., maeda, M. (1993). Measurements of thermally stratified pipe flow using image-processing techniques. Experiments in Fluids, Vol. 16, pp. 82-96.

Samy, R.; Glawdel, T.; Ren, C. L. (2008). Method for microfluidic whole-chip temperature measurement using thin-film poly (dimethylsiloxane)/rhodamine B. Analytical Chemistry, Vol. 80, pp. 369–375

Schneegass, I., Brautigam, R., kohler, J.M. (2001). Miniaturized flow-through PCR with different template types in a silicon chip thermocycler. Lab on a Chip, Vol. 1, pp. 42-49

Selva, B., Marchalot, J., Jullien, M.C., (2009). An optimized resistor pattern for temperature gradient control in microfluidics. Journal of Micromechanics and Microengineering, Vol. 19, 065002

Shah, J.,Geist, J., Locascio, L. E., Gaitan, M., Rao, M.V., Vreeland, W. N. (2006). Capillarity induced solvent-actuated bonding of polymeric microfluidic devices. Analytical Chemistry, Vol. 78, pp. 3348–3353

Shah, J. J., Sundaresan, S.G., Geist, J., Reyes, R.D., Booth, J.C. Rao. M. V., Gaitan, M. (2007). Microwave dielectric heating of fluids in an integrated microfluidic device. Journal of Micromechanics and Microengineering. Vol. 17, pp. 2224–2230

Shah. JJ. Microfluidic devices for forensic DNA Analysis. PhD thesis. George Mason University, Fairfax; 2007

Shah, J., G., Gaitan, M., Geist, J. (2009). Generalized temperature measurement equations for rhodamine B dye solution and its application to microfludics. Analytical Chemistry, Vol. 81, pp. 8260-8263

Shah, J., Geist, J., Gaitan, M. (2010). Microwave-induced adjustable nonlinear temperature gradients in microfludic devices. Journal of Micromechanics and Microengineering, Vol. 20, 105025

Sklavounos, A., Marchiarullo, D. J., Barker, S.L.R., Landers, J. P., barker, N. S. (2006). Efficient miniaturized systems for microwave heating on microdevices. Proceeding Micro Total Analysis Systems

Tanaka, Y., Slyadney, M.N., Hibara, A., Tokeshi, M., Kitamori, T. (2000). Non-contact photothermal control of enzyme reactions on a microchip by using a compact diode laser. Journal of. Chromatography.A, Vol. 894, pp. 45–51

Terry, S.C., Jerman, J.H., Angell, J.B. (1979). A gas chromatographic air analyzer fabricated on a silicon wafer. IEEE Transactions on Elecron Devices, Vol. 26, pp. 1880-1886

Tiercelin, N.; Coquet, P.; Senez, V.; Sauleau, R.; Fujita, H. (2006). Polydimethylsiloxane membranes for millimeter-wave planar ultra flexible antennas. Journal of Micromechanics and Microengineering, Vol. 16, pp. 2389-2395.

Whittaker, A.G., Mingos, D.M.P. (2002). Synthetic reactions using metal powders under microwave irradiation. Journal of the Chemical Society Vol. 21, pp. 3967-3970

Wei, Y. Z.; Sridhar, S. (1990). Journal of Chemical Physics, Vol. 92, pp. 923-928

Woolley, A.T., Hadley, D., Landre, P., deMello, A.J., Mathies, R.A., Northrup, M.A. (1996). Functional Integration of PCR Amplification and Capillary Electrophoresis in a Microfabricated DNA Analysis Device. Analytical Chemistry, Vol. 68, pp. 4081-4086

Yang, J.N., Liu, Y.J., Rauch, C.B., Stevens, R.L., Liu, R.H., Lenigk, R., Grodzinski, P., (2002). High sensitivity PCR assay in plastic micro reactors. Lab on a Chip, Vol. 2, pp. 179-187

Yoon, D.S., Lee, Y,S., Lee, Y., Cho, H.J., Sung, S,W., Oh, K.W., Cha, J., Lim, G. (2002). Precise temperature control and rapid thermal cycling in a micromachined DNA polymerase chain reaction chip. Journal of Micromechanic Microengineering, vol. 12, pp. 813-823

Zhang, H. D., Zhou, J., Xu, Z.R., Song, J., Dai, J., Fang, J., Fang, Z.L. (2007). DNA mutation detection with chip-based temperature gradient capillary electrophoresis using a slantwise radiative heating system. Lab on a Chip, Vol. 7, pp. 1162–1170

Chapter 8

Microwave Apparatus for Kinetic Studies and *in-situ* Observations in Hydrothermal or High-Pressure Ionic Liquid System

Masaru Watanabe, Xinhua Qi, Taku M. Aida and Richard Lee Smith, Jr.

Additional information is available at the end of the chapter

1. Introduction

1.1. Microwave heating on organic reactions

As many noticeable studies are introduced in this book, microwave heating technique has gained high expectation for utilizing various chemical processes including material synthesis, organic synthesis and conversion of energy resource with high reaction and energy efficiency. For organic synthesis, Gedye et al.[1] and Giguere et al.[2] prove effectiveness of microwave heating to accelerate organic reactions. Comparison of the energy efficiency between a conventional oil bath synthesis and a microwave-assisted synthesis has indicated that a significant energy savings of up to 85-fold can be expected using microwaves as an energy source on a laboratory scale[3]. It was also shown that microwave heating on organic chemical reaction had much higher yields within short reaction times for some products[4]. These high efficiency would be kept for a pilot scale plant and the plant must be greener chemical process because of pre-workup reduction and associated energy savings.

1.2. Importance of kinetic data and *in-situ* observation

To develop an industrial process for a new chemical reaction, one has to grasp a correct kinetics of the reaction as a fundamental data to design a reactor and set an operating condition. Kinetic data can be measured using either batch reactor, semi batch reactor and flow reactor. However, the important points to evaluate kinetics are to keep operating condition constant and to know reaction time correctly. Particularly, it is quite difficult for a batch and a semi-batch reactor to be achieved to a desired temperature rapidly and/or cooled down to enough low temperature spontaneously.

Furthermore, chemical reaction is affected with mass transfer occurring at phase boundary. Ionic liquid, which is widely investigated as reaction media for biomass conversion, is high viscous liquid depending on a solute concentration, amount of additive and temperature and a reaction in it is sometimes controlled by flow dynamics in a reactor. To know the effect of mass transfer on reaction kinetics, *in-situ* observation often provides a meaningful hint.

1.3. Concept for development of a new microwave heating apparatus

Advantage of microwave irradiation is capable of rapid heating *via* absorption of a heated target, mainly dielectric substance like water, carbon, some kinds of metal oxides, and so on. Now we consider microwave irradiation for high pressure vessel containing water. That is, to keep liquid phase of water over 100 °C, a pressure vessel has to be used to resist higher pressure than atmospheric one. In the high pressure vessel for microwave irradiation, rapid cool down is typically impossible due to lower heat transfer of a material (ceramics and plastics are used in a commercial set up) for the high pressure vessel. To cool down rapidly using air blow, which is a simple way for cooling, a metal material is favorable owing to its high heat conductivity (heat conductivity of stainless steel is around 20). But a metal material reflects microwave and it is not heated up by microwave irradiation. On the other hand, a microwave-transparent material has low heat conductivity (heat conductivity of ceramics is single digit and that of plastics is one digit smaller than ceramics) and rapid cooling by air blow can not be expected. In addition, high pressure vessel for microwave irradiation is composed of visible light-proof materials (which does not allow visible light pass through), for example, alumina, Teflon, and so on, and *in-situ* observation of reaction behavior in the vessel is basically impossible. To overcome these disadvantages, we developed a novel high-pressure reactor for microwave irradiation, which is capable of rapid heating, rapid cooling and *in-situ* observation. The key points of the reaction vessel are three as follows: (1) a commercial Pyrex glass cup and polycarbonate (PC) tube, of which transparency for visible light are quite high, are employed to compose the vessel as inner and outer tube, respectively, (2) heat insulating space between the inner glass and outer PC tube allow a reaction fluid in the glass cup to be heated up to 200 °C (PC is engineering plastics, however its heat tolerance is low and PC can normally be used up to 100 °C), and (3) the heat insulating space can be used for cooling unit after the reaction by introducing cooling water. The detail of the microwave setup and a typical procedure are describe below.

2. Microwave apparatus

2.1. Setup of microwave apparatus

Figure 1 shows a schematic diagram of the microwave apparatus. Figures 2 and 3 show cover shot of the setup and photograph of high pressure vessel (2 ~ 6 in Figure 1), respectively. The setup consists of a multimode microwave generator (1) μ-Reactor, SMW-087, 2.45 GHz, maximum power 700 W, Shikoku Keisoku, Takamtsu, Japan) with a K-type thermocouple (2), a stirring system (5 and 15), a control box (16:, a high pressure reactor (consists of 3, 4, 6 and 7), a pressure gauge (8), an inert gas (Ar or N_2) cylinder (13), a cooling water tank (18) and a vacuum pump (14). The reactor was composed of an inner thick-wall Pyrex glass tube (4: HPG-10, volume 10 ml, maximum supporting pressure 10 MPa, TaiatsuTechno. Corporation, Tokyo, Japan), an outer PC tube (3) and two PEEK (Teflon or PC is OK depending on operating temperature) screw caps (6 and 7) with special seal joint (consists of stainless steel connectors, Teflon O-ring, and Viton O-ring) used to fix glass tube and PC tube. The thermocouple (2) is inserted into the glass reactor (4) through a stainless steel sleeve and fixed with the inner wall of microwave oven to avoid microwave leakage from the oven (for this reason, an aluminum plate is also placed on a hole opened at the top of the microwave oven) and sparks produced from the thermocouple. Sparks from the thermocouple have never observed during all the experiments. The leakage of microwave was monitored by a microwave survey meter (Holiday Industries Inc., Model HI-1501) for safety and found to be less than 1 mW/cm^2 at distance of 5 cm far away from the microwave oven. The temperature inside the reactor is monitored and controlled by the control box (16). The temperature and power of microwave were monitored and recorded using a computer (21).

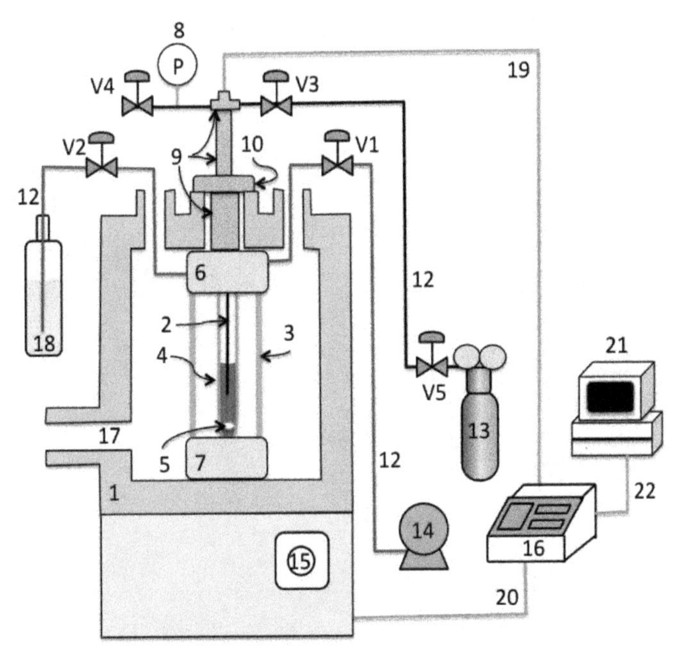

1- Microwave oven, 2- K-type Thermocouple 3- Polycarbonate outer tube, 4- Thick-walled glass reactor, 5- Stirrer bar, 6- PEEK, Tefron, or PC cap, 7- PEEK, Tefron, or PC bottom, 8- Pressure indicator, 9- Stainless steel connectors, 10- Aluminum plate, 11- PEEK line, 12- Stainless steel line, 13- inert gas (Ar or N_2) cylinder, 14- Vacuum pump, 15- Stirrer controller, 16- Power controller, 17- Observation window, 18- Cooling water tank, 19- Thermocouple connecting line, 20- Controller connecting line, 21- Computer, 22- Computer connecting line, V1~V5- stop valves

Figure 1. Microwave heating experimental setup

Figure 2. Cover shot of microwave heating experimental setup

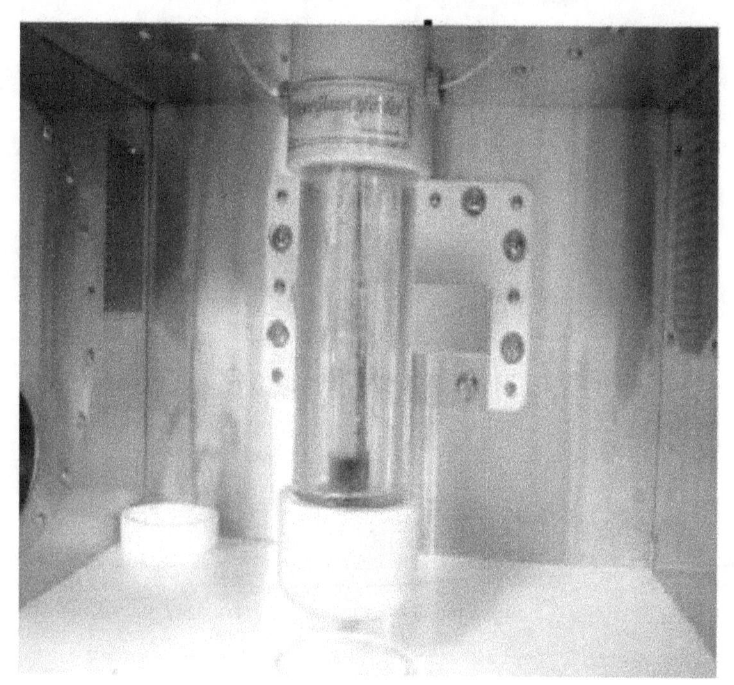

Figure 3. Photograph of high pressure vessel

2.2. Typical procedures for microwave heating experiments

A typical workup procedure for an experiment is as follows: An sample solution which must contain dielectric substance (water, polar solvents, ionic liquids, carbon, some kinds of metal oxides, etc) and sometimes a given amount of a catalyst are loaded into the glass tube (4) with a stirrer bar (5). The glass tube (4) is mounted into a PC tube (3) that is closed with PEEK screw caps (6 and 7: Teflon and PC caps are sometimes used depending on reaction temperature). This assembly is placed into the microwave oven (1) as shown in Figure 1. An inert gas (such as Ar and N_2) is used for purging air inside the reactor at a pressure of about 1.2 MPa. Then a vacuum pump (14) evacuates air from the space between the inner glass tube (4) and the outer PC tube (3) to minimize conductive heat losses. Introduction of cooling water into this vacuum space provided a method for rapidly cooling of the reactor. When microwave irradiation is started, the reaction mixture could be heated up to 200 °C within 60 s depending on the kind and amount of materials and substrates. Figure 4 shows a temperature and power of microwave profile at the experiments of fructose conversion in water in the presence of TiO_2 at 200 °C for 2 min 30 s. In this experiment, the reaction fluid was rapidly heated up to 200 °C for 30 s. After a desired reaction time passed, microwave irradiation was turned off, stop valve (V1) located at the line between the PC cap (6) and the vacuum pump (14) is closed, the vacuum pump (14) is stopped. After that, stop valve (V2) connected the line between the PC cap (6) and the cooling water tank (18) is opened to admit introduction of cooling water from tank (18) into the PC tube. By the cooling process, the reaction solution could rapidly be cooled down to below 80 °C within 60 s. As shown in Figure 4, at the fructose conversion experiment in water, the solution in the reaction vessel was rapidly cooled down to 80 °C within 30 s. After cooling, stop valve (V4) at the opposite site of the inert gas cylinder is opened and the inert gas inside the glass tube (4) is discharged. The reactor is disassembled and the reaction solution is collected by washing

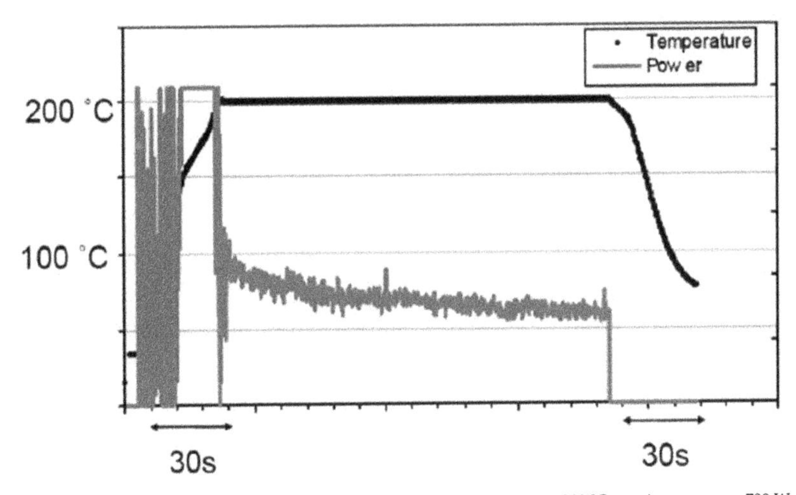

Reaction condition: 5 mL fructose aqueous solution, 0.05 g TiO_2, set temperature 200 °C, maximum power 700 W

Figure 4. Temperature and power profile during a reaction

the glass tube (4) with an amount of an appropriate solvent. During a reaction, reaction behavior can be observed from a observation window (17) and we sometimes record the images of some experiments with a digital video camera. One example is shown in the section of ionic liquid system concerning cellulose hydrolysis in an ionic liquid.

3. Experimental results and discussion

3.1. Topics described in this chapter

Special feature of microwave for improving reaction was noticed in organic synthesis for the first time. There have been many papers on microwave-assisted organic synthesis[1, 5-9] and only a limited number of studies have been reported about biomass conversion into a chemical block[10–13]. Here, we introduce four experimental results of biomass conversion in hydrothermal condition and ionic liquid systems. At first, fructose conversions into 5-hydroxylmethyl furfural (HMF), which is an important chemical block, in the presence of catalysts at hydrothermal condition[14] or in water mixture[15] are introduced. Second, we described partial oxidation of glycerol into formic acid with hydrogen peroxide at hydrothermal condition[16]. Both the hydrothermal reactions are described in the next section. The third topic is glucose transformation into HMF in an ionic liquid. We compared kinetics and product distribution of these reactions by microwave heating with those by a heating bath (outer heating such as a fluidized sand bath, a molten salt bath and an oil bath, which are capable of rapid heating as same heating rate as microwave heating). Finally, we demonstrated usefulness of *in-situ* observation of cellulose hydrolysis in high pressure ionic liquid-water mixture for considering importance of mass transfer (or phase separation) on the hydrolysis.

3.2. Experimental by outer heating

For the experiments using the outer heating, procedures for loading and recovering the samples were basically the same as those used in the microwave heating experiments, as explained in the previous section. Reactions were conducted with stainless steel 316 tube bomb reactors (inner volume: 6 mL). A given amount of solution (sample and solvent), catalyst (anatase TiO_2 for the fructose conversion at hydrothermal condition, Dowex 50wx8-100 ion-exchange resin for that in water-acetone mixture, $CrCl_3$ for glucose conversion in ionic liquid system), and/or additive (hydrogen peroxide for glycerol partial oxidation) were loaded into the reactor. Then, purging air inside the reactor was done by an inert gas (Ar or N_2) and the gas was fed in the reactor at 1.2 MPa of pressure. After the loading, the reactor was submerged into a heating bath and heated up to targeted temperature. The heating rate of the outer heating was as the same as the microwave heating (within 90 s for either heating bathes employed in this study). Here, sands fluidized in the fluidizing sand bath are fused alumina particles and the salt in the molten salt bath is 50 wt% KNO_3-50wt% $NaNO_3$ salt mixture. Silicone oil was used for an oil bath used in the study on glucose conversion in an ionic liquid. After the reaction time was passed, the reactor was taken out of the bath and quenched in a water bath that was at room temperature. The cooling rate in the water bath of the stainless steel reactor was also the same as that by the cooling system in the microwave setup. Before opening the reactor, the gas in the reactor was released. Even for the partial oxidation experiments, gas analysis was not conducted. The liquid samples in the reactor were collected with rinsing of the reactor with water. The solid in the recovered solution, namely anatase TiO_2 or Dowex 50wx8-100 for the fructose reaction, was separated from the liquid sample by filtration before analysis. No solid was obtained at the partial oxidation of glycerol and the HMF formation from glucose

3.3. Fructose conversion into HMF at hydrothermal condition[14]

Firstly, the effect of microwave heating on the fructose conversion in hydrothermal condition[15] was mentioned. The experimental results are listed in Table 1 and parity plot of these values are plotted in Figure 5. The major products obtained by both the outer heating and the microwave heating, were glucose, HMF, furfural and organic acids (lactic acid, hydroxyacetone, formic acid, acetic acid and levulinic acid). The product yields for the experiments by microwave heating were totally higher than those by the outer heating. Especially, the most dominating product, HMF, was obtained much more for the microwave heating than for the outer heating. As shown in Table 1, for 3 min reaction time, the fructose conversion and HMF yields for the outer heating bath were 35 mol% and 12 mol%, respectively, while the corresponding values for the microwave heating were 73 mol% and 27 mol%, respectively. For 5 min of the reaction, the fructose conversion (65 mo% by the outer heating and 84 mol% by the microwave heating) and the HMF yield (27 mo% by the outer heating and 34 mol% by the microwave heating) were also enhanced. Both fructose conversions and HMF yields were enhanced with microwave heating compared with the outer heating, while the HMF selectivities obtained by the outer heating (34 % for 3 min and 41 % for 5 min) was almost the same as those by the microwave heating (38 % for 3 min and 40 % for 5 min). The parity plots of the reactions in the outer heating and the microwave heating clearly show the differences. Figure 5 apparently indicates that the microwave heating enhanced the total conversion of fructose but the reaction of fructose into HMF was not strongly affected by the heating method. It was well known that HMF was formed via dehydration from fructose and further reacted with water to form levulinic acid. At an early stage of the fructose reaction (namely shorter reaction time), HMF formation was dominantly occurred but the produced HMF was degraded into the other products at a latter stage of the reaction. As indicated in Table 1 and Figure 5, the HMF selectivity at the microwave heating for 3 min was a little higher than that at the outer heating. This probably indicated that the reaction pathway of HMF formation was slightly promoted by the microwave heating.

	Time, min	Fructose conversion,mol %	HMF yield, mol %	HMF selectivity, %
Outer heating	3	35.3	12.1	34.3
	5	65.3	26.9	41.2
Microwave heating	3	73.1	27.9	38.2
	5	84.1	33.5	39.8

Condition: 5 mL of 2wt% of fructose, 0.2g of anatase TiO_2, 200 °C of reaction time

Table 1. Comparison of fructose conversion between the outer heating and microwave heating[15]

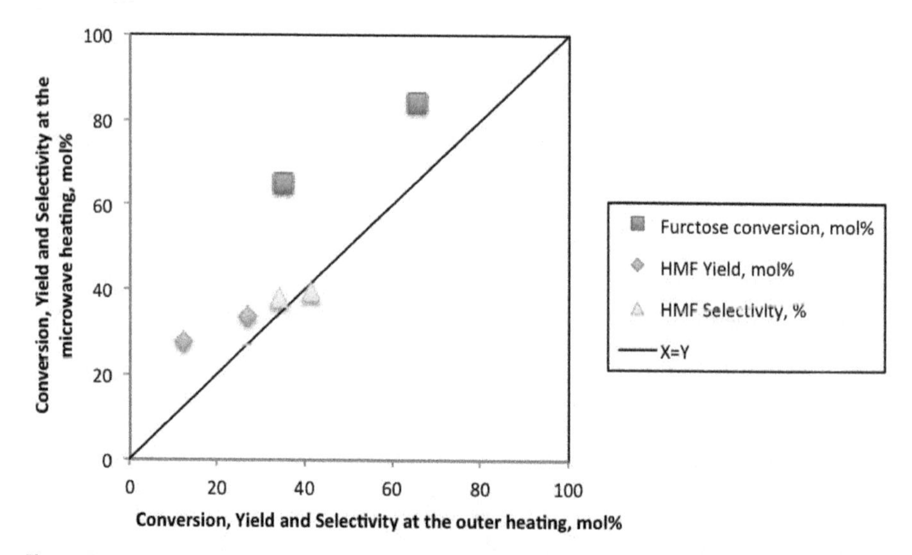

Figure 5. Parity plot of the fructose conversion at hydrothermal condition: comparison of fructose conversion, HMF yield and HMF selectivity between the outer heating and the microwave heating

3.4. Fructose conversion into HMF in acetone-water mixture[15]

The dehydration of D-fructose (2 wt%) in an acetone–water mixture (70 : 30, w/w) with an ion-exchange resin as catalyst by convective heating (sand bath) and microwave irradiation heating was also studied at 150 °C with keeping liquid phase by giving over the vapor pressure (Table 2 and Figure 6). Figure 6 is also parity plot as well as Figure 5. That is, when a plot in the graph is higher than X=Y line, the phenomenon is favored to be occurred by the microwave heating. As shown in Figure 6, microwave irradiation was significantly more efficient not only for fructose conversion but also for 5-HMF yields. In addition, the selectivity of HMF formation by the microwave heating was slightly higher that by the outer heating.

	Reaction time, min	Fructose conversion, mol%	HMF Yield, mol%	HMF Selectivity, %
Outer heating	5	7.8	5.7	73.1
	10	22.1	13.9	62.9
Microwave heating	5	64.5	54.0	83.7
	10	91.7	70.3	76.7

Condition: acetone:water=70:30 w/w, 5mL of 2 wt% fructose solution,0.1 g Dowex 50wx8-100, 150°C of reaction temperature

Table 2. Comparison of fructose conversion between the outer heating and microwave heating[16]

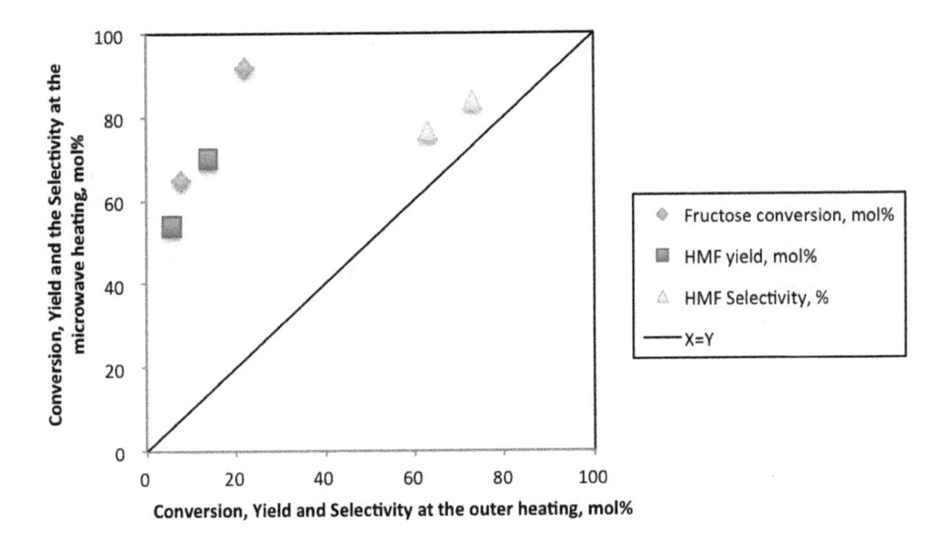

Figure 6. Parity plot of the fructose conversion in acetone-water mixture: comparison of fructose conversion, HMF yield and HMF selectivity between the outer heating and the microwave heating

3.5. Partial oxidation of glycerol into formic acid at hydrothermal condition[16]

Figure 7 shows the experimental results of partial oxidation of glycerol in hydrothermal condition[16]. The reaction condition was 0.5 of ER, 200 °C of reaction temperature, 10 min of reaction time. Here, ER is equivalent ratio of oxygen and is defined as below:

$$ER[-] = \frac{\textit{Loaded}\text{ amount of oxygen atom } [mol]}{\textit{Required}\text{ amount of oxygen atom for complete oxidation } [mol]} \quad (1)$$

1 mole of glycerol ($C_3H_8O_3$) was completely oxidized by 7/2 moles of O_2 into 3 moles of CO_2 and 4 moles of H_2O. 1 mole of hydrogen peroxide is decomposed into 1 mole of H_2O and 1/2 moles of O_2 and so the ratio of glycerol to hydrogen peroxide in the batch type reactor (high pressure glass reactor for the microwave heating and stainless steel reactor for the outer heating) was 1:3.5. Here we focused on formic acid formation from glycerol because formic acid was expected as a sustainable hydrogen storage resource because it can easily decompose into hydrogen when needed[16].

As shown in Figure 7, glycerol conversion and formic acid yield obtained by the microwave heating was a little higher than those by the outer heating. For the selectivity of formic acid formation, the microwave heating also enhanced.

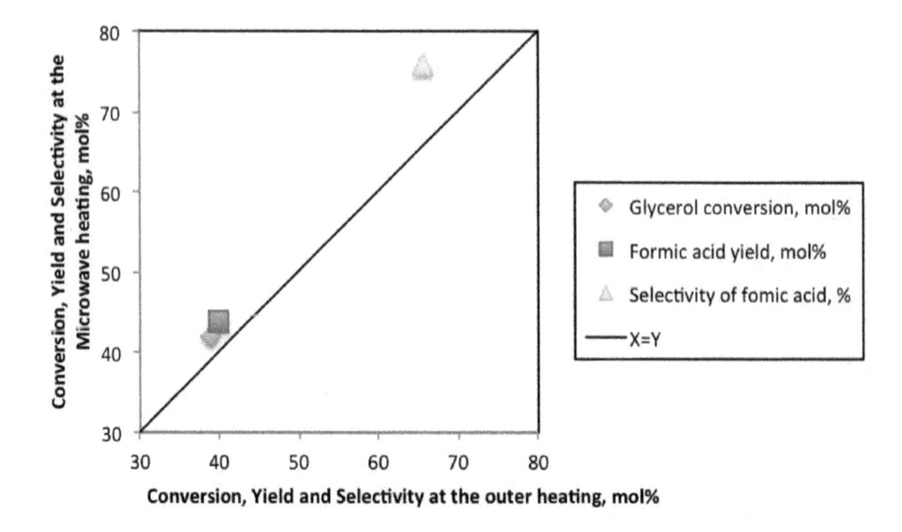

Figure 7. Parity plot of the glycerol partial oxidation: comparison of heating method for glycerol partial oxidation (200 °C, 10 min, ER = (oxygen atom in the reactor)/(oxygen molecular atom for complete oxidation) = 0.5)

3.6. Glucose conversion into HMF in[Bmim][Cl][17]

In ionic liquid, 1-methyl-3-butyl imidazolium chloride,[Bmim][Cl], glucose conversion into HMF in the presence of $CrCl_3$ was carried out with outer heating (oil bath heating) and microwave heating at identical conditions for two temperatures, shown in Table 3. To compare the differences between the outer heating and the microwave heating, the experimental data are plotted in Figure 8 as parity plot. It was apparently seen that both glucose conversion and the reaction pathway to HMF formations were promoted some degrees by microwave irradiation.

	Reaction temperature, °C	Reaction time, min	Glucose conversion,mol%	HMF yield, mol%	HMF selectivity, %
Outer heating	120	5	63	45	71.4
	140	0.5	68	48	70.6
Microwave heating	120	5	89	67	75.3
	140	0.5	96	71	74.0

Condition: 1 g of 1wt% of glucose in[Bmim][Cl], 0.015 g $CrCl_3 \cdot 6H_2O$

Table 3. Comparison of glucose conversion between the outer heating and microwave heating[18]

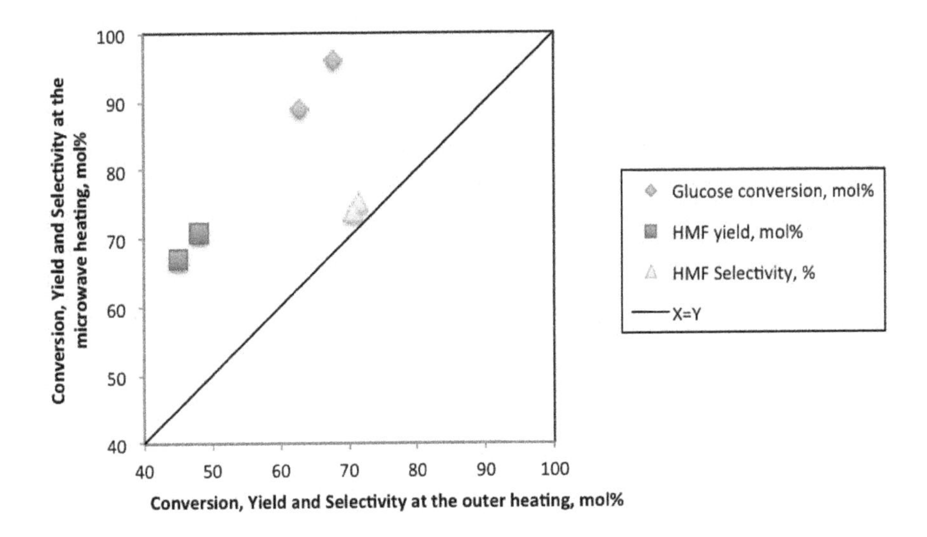

Figure 8. Parity plot of the glucose conversion: comparison of glucose conversion, HMF yield and HMF selectivity between the outer heating and the microwave heating

3.7. Effect of microwave heating on the biomass conversions

As introduced the above sections, the enhancement of reaction rate and some degrees of change of reaction pathway were seen for glucose, glycerol and fructose conversion in water and ionic liquid system. Microwave effect has been considered as thermal and/or specific effect.

Lidstroem et al. reviewed microwave-assisted organic syntheis[18]. In their review, they mentioned that some organic reactions were speeded up by microwave heating and the enhancement of the chemical reactions was mainly resulted in the difference of temperature between conductive heating (heat transferred from outer heating source) and inner heating (heat .is directly supplied by microwave heating). Conventionally, a typical organic synthesis is conducted with glassware and heat transfer from outer heater to a reaction fluid in the glass is slow. On the other hand, microwave heating is capable of rapid heating because of direct energy supply to the reaction fluid. Also, microwave is sometimes absorbed heterogeneously in a reaction vessel and hot spot (regionally higher temperature place exist in the reaction system) is probably encountered. The difference of temperature profile and/or the existence of the hot spot would affect chemical reactions and the researcher considered that microwave has some specific effect on some reactions. At the experimental studies described in this chapter, the heating profiles and the final temperature for both the reactions by the outer heating and the microwave heating were tried to be identical and thus the effect of the achieved temperature and the heating rate on the reaction could be excluded from the reason of the difference of the biomass conversions. For the viewpoint of hot spot, it is difficult to know whether hot spot exit or not, however the explanation of the reason why the reaction behavior was changed by the microwave heating is difficult. At all of the biomass conversion introduced in this chapter, selectivity of the product was also enhanced and it means that microwave selectively enhanced a specific reaction pathway. This promotion of specific reaction can not be explained from the thermal effect.

It has been known that some difficult-to-rationalize effects (which are referred as specific or nonthermal effects) were seen in some organic reactions[6]. In general, specific effect of microwave has been proposed to be the result of a direct interaction of dipoles in an electric field with a specific functional group in the reaction medium. It has been argued that there is a decrease in activation energy1[9] or an increase in the pre-exponential factor in the Arrhenius law due to enhanced orientation effects of polar species in an electromagnetic field that lower steric hindrances[5, 20]. Some researchers have proposed that the effective collision among reactant molecules under microwave irradiation is enhanced and result in the pre-exponential factor in the Arrhenius law increases[5]. Furthermore, a similar effect should be observed for chemical reaction between reactants containing polar groups, where the rotation of polar species increases and the molecules effectively leads to activation (from ground state to transition state) under the microwave irradiation, thus enhance the reactivity by lowering the activation energy[5]. Bren et al.[21] proposed a novel physical mechanism for microwave catalysis based on rotationally excited reactive species and verify its validity through a computer simulation of a realistic chemical reaction. They thought that the rotation rate of polar molecules are enhanced under the microwave irradiation, this gives a higher rotational temperature than the translational temperature. The activation free energy is reduced when the rotational temperature is higher than the translational temperature, which constitutes a catalytic effect. These specific effects of microwave heating on organic reactions including thermal effect like hot spot and molecule-level phenomenon and it is difficult to be clarified experimentally. To reveal the effect of "specific effect" of microwave irradiation molecular level studies such as quantum chemical calculations is strongly required.

3.8. *In-situ* observation of cellulose hydrolysis in ionic liquid

Finally, in this section, we describe usefulness of the microwave setup developed by us for *in-situ* observation. Here, cellulose hydrolysis in ionic liquid ([Bmim][Cl]) is picked up. Figure 9 shows effect of loading amount of reaction fluid (5 wt% cellulose ionic liquid solution) in the high pressure glass reactor (4 in Figure 1) on cellulose hydrolysis. When cellulose hydrolysis progresses stoichiometrically, 3.1×10^{-4} mole (which is equal to glucose unit in 0.05g of cellulose) of water is required in the cease of 1 g of the amount of the solution. The water content in the solution in[Bmim][Cl] that was used in this study was 0.8 wt% (4.4×10^{-4} mole) and it was enough amount for cellulose hydrolysis. Reaction temperature was 120 ˚C and reaction time was 20 min. To keep water liquid phase, 1.2 MPa of inert gas was loaded before heating up to 120 ˚C. Catalyst was Amberlyst-15 and its amount was equal to the loaded amount of cellulose (thus when 1 g of the solution was loaded in the reactor, 0.05 g of Amberlyst-15 was loaded). As shown in this figure, with increasing the amount of the solution, glucose yield gradually decreased. More than 3g of the solution in the reactor, glucose yield changed unsteadily with increasing the amount of the solution. Figure 10 shows *in-situ* observation of cellulose hydrolysis when 3 g of the solution was loaded. As shown in Figure 10, ionic liquid ([Bmim][Cl]), cellulose, and water were miscible at first (Figure 10-1). With progressing hydrolysis of cellulose, phase separation was observed (Figure 10-2). Further progression of the reaction, water-rich phase (upper phase in the reactor) was completely separated from ionic liquid-rich phase (lower phase) and cellulose layer was also observed between the upper and lower phase (Figure 10-3).

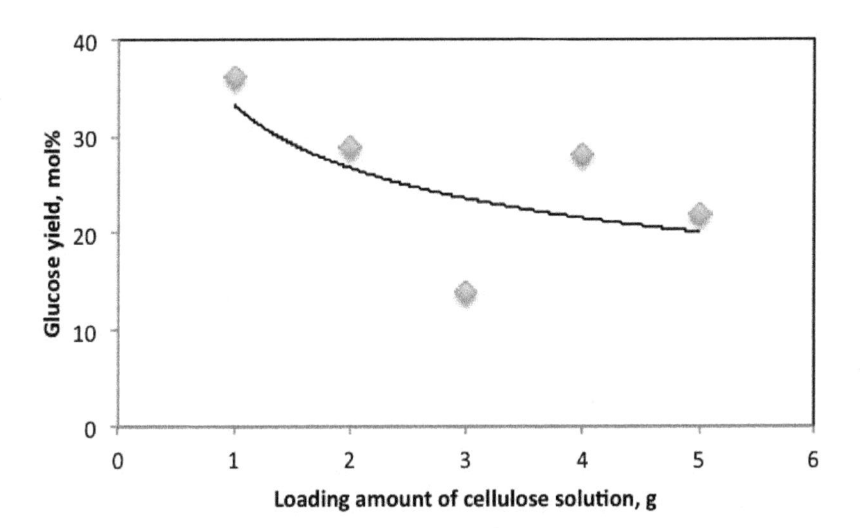

Condition: 5 wt% of cellulose, Cellulose/Amberlyst = 1, 120 ˚C, 20 min

Figure 9. Effect of loading amount of cellulose solution on glucose yield from cellulose hydrolysis in[Bmim][Cl] in the presence of Amberlyst-15

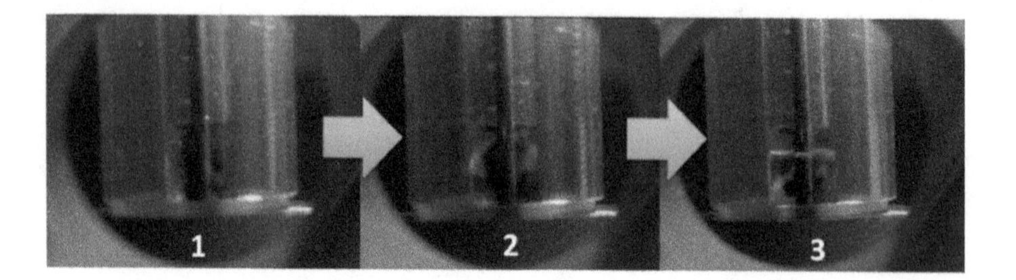

1. Ionic liquid, cellulose, and water were firstly miscible.
2. With progressing hydrolysis of cellulose, phase separation was observed.
3. Water-rich phase was completely separated from ionic liquid-rich phase and cellulose layer was also observed.

Figure 10. *In-situ* observation of cellulose hydrolysis in ionic liquid and water mixture.

Viscosity of ionic liquid is high, particularly when ionic liquid contains cellulose and the viscosity of the ionic liquid solution is much higher than pure ionic liquid. During the hydrolysis, the viscosity of the reaction fluid gradually decreases with increasing product (mainly glucose) yield. When stirring in the reactor was insufficiency, viscosity and density of the solution was disproportionate. In addition, the solid catalyst was used and it was aggregated when the viscosity (and density) was heterogeneously distributed. The phase boundary formed with unreacted cellulose prohibited mass transfer and the disproportion of the solution was accelerated. Surely, although there are several reasons for explaining the reaction behavior shown in Figure 9, the effect of mass transfer (also phase behavior) has to be the main factor for reducing glucose yield.

In this chapter, we only introduce one example for effectiveness of *in-situ* observation but *in-situ* observation definitely provides a useful hint for considering kinetics of many reactions. Our microwave system employs both microwave- and visible light-transparent material for high pressure reactor and whole the reactor can be observed during the reaction.

4. Conclusion

A novel high pressure reactor was introduced in this chapter. The apparatus allows kinetic study because of rapid heating and rapid cooling. Also *in-situ* observation inside a reactor during reaction is possible.

Here, dehydration of fructose into hydroxymethyl furfral (it is a chemical block made from lignocellulose material for biomass refinery), glycerol into formic acid, and glucose into HMF were mentioned. All of the reactions were affected by microwave irradiation. Through the experimental results, we tried to discuss the effect of microwave irradiation.

The usefulness of *in-situ* observation for cellulose hydrolysis in[Bmim][Cl] was shown. During the reaction, phase was changed from homogeneous into heterogeneous and it was considered that the phase change mainly resulted in inhibition of cellulose hydrolysis.

Author details

Masaru Watanabe, Taku M. Aida and Richard Lee Smith, Jr.
Tohoku University, Japan

Xinhua Qi,
Nankai University, China

5. References

[1] R. Gedye, F. Smith, K. Westaway, H. Ali, L. Baldisera, L. Laberge, J. Rousell, Tetrahedron Lett. 27 (1986) 279.
[2] R.J. Giguere, T.L. Bray, S.M. Duncan, G. Majetich, Tetrahedron Lett. 27 (1986) 4945.
[3] M.J. Gronnow, R.J. White, J.H. Clark, D.J. Macquarrie, Org. Process. Res. Dev. 9 (2005) 516.
[4] D. Dallinger, C.O. Kappe, Chem. Rev. 107 (2007)
[5] M. Hosseini, N. Stiasni, V. Barbieri, C.O. Kappe, J. Org. Chem. 72 (2007) 1417.
[6] C.O. Kappe, Angew. Chem. Int. Edit. 43 (2004) 6250.
[7] M. Kremsner, A. Stadler, C.O. Kappe, Microwave Meth. Organ. Synth. 266 (2006) 233.
[8] E. Comer, M.G. Organ, J. Am. Chem. Soc. 127 (2005) 8160.
[9] G. Shore, S. Morin, M.G. Organ, Angew. Chem. Int. Edit. 45 (2006) 2761.
[10] C. Limousin, J. Cleophax, A. Petit, A. Loupy, G. Lukacs, J. Carbohydr. Chem. 16 (1997) 327.
[11] A. Orozco, M. Ahmad, D. Rooney, G. Walker, Process. Saf. Environ. 85 (2007) 446.
[12] A.M. Sarotti, R.A. Spanevello, A.G. Suarez, Green Chem. 9 (2007) 1137.
[13] M.M. Andrade, M.T. Barros, P. Rodrigues, Eur. J. Org. Chem. (2007) 3655.
[14] X. Qi, M. Watanabe, T. M. Aida, R. L. Smith, Jr., Catal. Commun. 9 (2008) 2244
[15] X. Qi, M. Watanabe, Taku M. Aida, R. L. Smith, Jr., Green Chem., 10 (2008) 799.
[16] M. Watanabe, T. M. Aida, R. L. Smith Jr., H. Inomata, J. Jpn. Petrol. Inst., submitted.
[17] X. Qi, M. Watanabe, T. M. Aida, R. L. Smith, Jr., ChemSusChem 3 (2010) 3.
[18] P. Lidstroem, J. Tierney, B. Wathey, J. Westman, Tetrahedron, 57 (2001) 9225.
[19] F. Joo, ChemSusChem 1 (2008) 805.

[20] C.O. Kappe, A. Stadler, Microwaves in Organic and Medicinal Chemistry, Wiley-VCH, Weinheim, Germany, 2005.
[21] U. Bren, A. Krzan, J. Mavri, J. Phys. Chem. A 112 (2008) 166.

Permissions

The contributors of this book come from diverse backgrounds, making this book a truly international effort. This book will bring forth new frontiers with its revolutionizing research information and detailed analysis of the nascent developments around the world.

We would like to thank Wenbin Cao, for lending his expertise to make the book truly unique. He has played a crucial role in the development of this book. Without his invaluable contribution this book wouldn't have been possible. He has made vital efforts to compile up to date information on the varied aspects of this subject to make this book a valuable addition to the collection of many professionals and students.

This book was conceptualized with the vision of imparting up-to-date information and advanced data in this field. To ensure the same, a matchless editorial board was set up. Every individual on the board went through rigorous rounds of assessment to prove their worth. After which they invested a large part of their time researching and compiling the most relevant data for our readers. Conferences and sessions were held from time to time between the editorial board and the contributing authors to present the data in the most comprehensible form. The editorial team has worked tirelessly to provide valuable and valid information to help people across the globe.

Every chapter published in this book has been scrutinized by our experts. Their significance has been extensively debated. The topics covered herein carry significant findings which will fuel the growth of the discipline. They may even be implemented as practical applications or may be referred to as a beginning point for another development. Chapters in this book were first published by InTech; hereby published with permission under the Creative Commons Attribution License or equivalent.

The editorial board has been involved in producing this book since its inception. They have spent rigorous hours researching and exploring the diverse topics which have resulted in the successful publishing of this book. They have passed on their knowledge of decades through this book. To expedite this challenging task, the publisher supported the team at every step. A small team of assistant editors was also appointed to further simplify the editing procedure and attain best results for the readers.

Our editorial team has been hand-picked from every corner of the world. Their multi-ethnicity adds dynamic inputs to the discussions which result in innovative outcomes. These outcomes are then further discussed with the researchers and contributors who give their valuable feedback and opinion regarding the same. The feedback is then collaborated with the researches and they are edited in a comprehensive manner to aid the understanding of the subject.

Apart from the editorial board, the designing team has also invested a significant amount of their time in understanding the subject and creating the most relevant covers. They scrutinized every image to scout for the most suitable representation of the subject and create an appropriate cover for the book.

The publishing team has been involved in this book since its early stages. They were actively engaged in every process, be it collecting the data, connecting with the contributors or procuring relevant information. The team has been an ardent support to the editorial, designing and production team. Their endless efforts to recruit the best for this project, has resulted in the accomplishment of this book. They are a veteran in the field of academics and their pool of knowledge is as vast as their experience in printing. Their expertise and guidance has proved useful at every step. Their uncompromising quality standards have made this book an exceptional effort. Their encouragement from time to time has been an inspiration for everyone.

The publisher and the editorial board hope that this book will prove to be a valuable piece of knowledge for researchers, students, practitioners and scholars across the globe.

List of Contributors

S.M. Javad Koleini
Tarbiat Modares University, Iran

Kianoush Barani
Lorestan University, Iran

Graham Brodie
Melbourne School of Land and Environment, University of Melbourne, Australia

Mohamed S. Shaheen, Khaled F. El-Massry and Ahmed H. El-Ghorab
Flavour and Aroma Department, National Research Center, Egypt

Faqir M. Anjum
National Science & Technology (NIFSAT), Agriculture University, Faisalabad, Pakistan

G.E. Ibrahim, A.H. El-Ghorab, K.F. El-Massry and F. Osman
Chemistry of Flavour and Aroma Department, National Research Center, Tahrir St. Dokki, Cairo, Giza, Egypt

Boris I. Kharisov, Oxana V. Kharissova and Ubaldo Ortiz Méndez
Universidad Autónoma de Nuevo León, Monterrey, México

F. Osada
Analysis Solution Engineering Section, 2nd Engineering Department Higashimurayama Plant Industrial Division, NIKKISO, LTD., Japan

T. Yoshioka
Graduate School of Environmental Studies, Tohoku University

Mulpuri V.Rao
Department of Electrical and Computer Engineering, George Mason University, Fairfax, VA, USA

Jayna J. Shah
National Institute of Standards and Technology, Semiconductor Electronics Division, Gaithersburg, MD, USA
Presently at: Raytheon Company, Dallas, TX, USA

Jon Geist and Michael Gaitan
National Institute of Standards and Technology, Semiconductor Electronics Division, Gaithersburg, MD, USA

Masaru Watanabe, Taku M. Aida and Richard Lee Smith Jr.
Tohoku University, Japan

Xinhua Qi
Nankai University, China